AF762946

Tools for High Performance Computing 2009

Matthias S. Müller • Michael M. Resch •
Alexander Schulz • Wolfgang E. Nagel
Editors

Tools for High Performance Computing 2009

Proceedings of the 3rd International Workshop on Parallel Tools for High Performance Computing, September 2009, ZIH, Dresden

Springer

Editors

Michael M. Resch
Alexander Schulz
Höchstleistungsrechenzentrum
Stuttgart (HLRS)
Universität Stuttgart
Nobelstraße 19
70569 Stuttgart
Germany
resch@hlrs.de
schulz@hlrs.de

Matthias S. Müller
Wolfgang E. Nagel
Zentrum für Informationsdienste
und Hochleistungsrechnen (ZIH)
Technische Universität Dresden
01062 Dresden
Germany
matthias.mueller@tu-dresden.de
wolfgang.nagel@tu-dresden.de

Front cover figure: Implicit representation and adaptive mesh of the scala tympani in the chochlea of a minipig.

ISBN 978-3-642-11260-7 e-ISBN 978-3-642-11261-4
DOI 10.1007/978-3-642-11261-4
Springer Heidelberg Dordrecht London New York

Library of Congress Control Number: 2010927589

Mathematics Subject Classification (2010): 68-06, 68Q85, 68Q60, 68N19, 68U99, 94A99, 68N99

© Springer-Verlag Berlin Heidelberg 2010
This work is subject to copyright. All rights are reserved, whether the whole or part of the material is concerned, specifically the rights of translation, reprinting, reuse of illustrations, recitation, broadcasting, reproduction on microfilm or in any other way, and storage in data banks. Duplication of this publication or parts thereof is permitted only under the provisions of the German Copyright Law of September 9, 1965, in its current version, and permission for use must always be obtained from Springer. Violations are liable to prosecution under the German Copyright Law.
The use of general descriptive names, registered names, trademarks, etc. in this publication does not imply, even in the absence of a specific statement, that such names are exempt from the relevant protective laws and regulations and therefore free for general use.

Cover design: WMXDesign GmbH, Heidelberg

Printed on acid-free paper

Springer is part of Springer Science+Business Media (www.springer.com)

Preface

Since its beginning, parallel computing has been a challenge, where new programming paradigms, operating systems, libraries and applications had to be developed to harvest the potential of multi-processor machines. The development of MPI and OpenMP provided basic programming support and a range of tools were developed to keep up with the hardware development.

Over the last years the landscape has changed. With the advent of multi-core CPUs parallelism has reached a new level. Tools and applications have to change both their pace of development and their scope. Now even the smallest computing devices are based on multi-core technology and require parallel programming. On the high end, the dramatic increase of parallelism has accelerated the growth of performance even well beyond Moore's law. As a consequence there is an increasing need for tools supporting the development of parallel applications. Constant progress and research is required to achieve that goal.

Support for such development has to supplement the investments made for HPC hardware. Since parallelism is a pervasive technology, development of parallel methods, tools and applications can now easily be brought to a much wider market. As a consequence, the Federal Ministry of Science in Germany has launched a funding initiative for development of software in HPC. Other countries – specifically the US and Japan – have launched similar programs recently. Further activities both at the national and at the European level will have to follow to turn this into a both fruitful and sustainable research and development environment.

The Parallel Tools Workshop that took place in Dresden on Sep. 14/15, 2009 was the third in a series of workshops that started 2007 at the High Performance Computing Center Stuttgart (HLRS). The goal of this series is to bring together tool developers and users from science and industry in an interactive environment. Participants from research and developers from science and industry were invited to discuss the most recent development, risk and opportunities and to exchange their ideas. The very interactive workshop attracted about 90 scientists from all over the world. The focus was on presenting a wide range of technologies, but also on giving hands-on sessions to demonstrate the strengths of each tool. The most interesting contributions are presented in papers in this collection of ideas, that continues the series of workshop proceedings for tools in high performance computing.

This year's presentations have been in the fields of Integrated Development Environments, Parallel Debugging and Performance Analysis tools from a wide range of scientific and industrial tool developers. This includes tools from vendors such as Allinea, Intel, Sun, and Totalview, as well as research institutions, including the University of Oregon, University of Houston, Iowa State University, Munich University of Technology, Ludwig-Maxmimilians-Universität München, University of Tennessee, University of Utah and the University of California. Contribution of research and computer centers came from Research Center Juelich, Barcelona Supercomputing Center, Oak Ridge National Laboratory, Lawrence Livermoore National Laboratory and the Center for Information Services and High Performance Computing.

We would like to acknowledge the support of Blasius Czink, Shiqing Fan, José Gracia and Christoph Niethammer.

Dresden, September 2009

Matthias Müller, Michael Resch
Alexander Schulz, Wolfgang Nagel

Contents

List of Contributors

Chapter 1
PERISCOPE: An Online-Based Distributed Performance Analysis Tool

Shajulin Benedict, Ventsislav Petkov, and Michael Gerndt

Abstract This paper presents PERISCOPE - an online distributed performance analysis tool that searches for a wide range of performance bottlenecks in parallel applications. It consists of a set of agents that capture and analyze application and hardware-related properties in an autonomous fashion. The paper focuses on the Periscope design, the different search methodologies, and the steps involved to do an online performance analysis. A new graphical user-friendly interface based on Eclipse is introduced. Through the use of this new easy-to-use graphical interface, remote execution, selection of the type of analysis, and the inspection of the found properties can be performed in an intuitive and easy way. In addition, a real-world application, namely, the GENE code, a grand challenge problem of plasma physics is analyzed using Periscope. The results are illustrated in terms of found properties and scalability issues.

1.1 Introduction

Performance is one of the key concerns for the application developers and the computational resource providers of the scientific community. The application developers endeavor writing efficient large-scale scientific codes that make optimal use of supercomputers or any other computational resource. Very often, they fail owing to the required knowledge about the architectures, memory hierarchy, and networks connecting the processors. The factors which determine a program's performance, such as, memory usage, I/O, compilers, operating system and so forth, are complex, inter-related, and frequently hidden from the programmer.

Performance analysis tools help users in writing efficient codes for current High Performance Computing (HPC) machines. These tools can provide the user with

Shajulin Benedict, Ventsislav Petkov and Michael Gerndt
Technische Universität München
Fakultät für Informatik I10, Boltzmannstr. 3, 85748 Garching, Germany

M.S. Müller et al. (eds.), *Tools for High Performance Computing 2009*,
DOI 10.1007/978-3-642-11261-4_1, © Springer-Verlag Berlin Heidelberg 2010

measurements of the program's performance and locate various bottlenecks. Bottlenecks are places in the execution path where execution time is lost due to inefficient resource usage. Based on the identified bottlenecks, users can do modifications to improve the application's runtime behavior. Since measuring performance data and storing those data for further analysis is often done in a not very scalable approach, most tools are limited to experiments with a small number of processors.

The traditional way of conducting performance analysis and tuning for high performance computing has been an off-line search approach requiring strong involvement of the user. This search has a potential problem with large performance datasets and long analysis times for large-scale scientific applications. It remains a challenge for application developers to analyze the bottlenecks of their applications when scaling to larger parallel machines. To investigate the runtime behavior of large experiments, performance analysis has to be done online in a distributed fashion, eliminating the need to transport huge amounts of performance data through the parallel machine's network and to store those data in files for further analysis.

An online-based performance analysis system using expert knowledge for identifying bottlenecks in the applications, in general, follows four steps for capturing performance properties (Fig. 1.1). As shown, the application is instrumented based

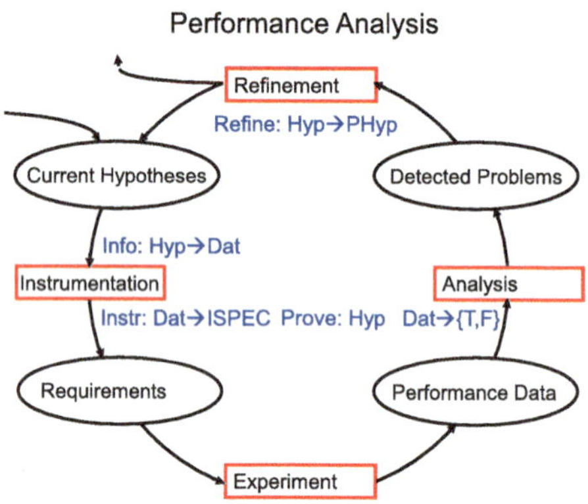

Fig. 1.1: Cyclic representation of performance analysis

on the initial hypotheses of potential performance properties. During an experiment executing the application on the parallel system, appropriate performance data are collected. These data are then inspected to prove which of the hypotheses hold. In the refinement step, the found properties might be refined to identify more specific performance problems. All four steps are executed in a cyclic fashion until no more precise properties can be found. This cyclic approach can of course be automated and executed in an online fashion.

Although, there are numerous performance analysis tools on the market, they face challenges in usabiliy, scalability, and single node performance analysis. Periscope [5] is a distributed online performance analysis tool currently under development at Technische Universität München that addresses the above mentioned challenges. It consists of a set of autonomous agents that search for performance properties. Each agent is responsible for analyzing a subset of the application's processes and threads. The agents request measurements from the monitoring system, retrieve the data, and use it to identify performance properties.

Periscope is currently extended within the German project ISAR (Integrated System and Application Analysis for Massively Parallel Computers in the Petascale Range[1]) funded until 2011. The main goal of the ISAR project is the realization of an integrated scalable system that can be used in production environments.

This paper gives a short overview of Periscope's design. It also presents for the first time the usage model of Periscope which is based on the recent work within the ISAR project. It introduces the new graphical user interface which facilitates the user's interaction with Periscope. In addition, we give some results from a large-scale run of a real-world application, namely, the Gyrokinetic Electromagnetic Numerical Experimental code (GENE) from the Max Planck Institute for Plasma Physics in Garching.

All the experiments were done on our Altix 4700 at Leibniz Rechenzentrum (LRZ) in Garching. The Altix supercomputer consists of 19 NUMA-link4 interconnected shared memory partitions with over 9600 Itanium 2 cores with an aggregated peak performance of over 60 TFlops.

The paper is organized as follows: Section 1.2 introduces related work and Section 1.3 describes the design of Periscope with an detailed overview of the analysis agent's data capturing mechanism. The usage model is presented in Section 1.4 and the distributed performance analysis is illustrated in Section 1.5.

1.2 Related Work

Just as there are several different classes of parallel hardware and parallel programming languages, so too are there several distinct types of performance analysis tools. Each of them has a number of concreate realizations. The most notable ones are Paradyn, TAU, Vampir, KOJAK, SCALASCA, and mpiP.

Paradyn [9, 10], developed during 1990s, automates performance analysis. The toolkit uses a performance consultant that does dynamic instrumentation and searches for bottlenecks based on summary information during the program's execution. The tool has the capability to scale to large numbers of processors.

TAU [12] does offline analysis of performance issues in the application based on trace-based and profiling-based approaches. It provides a wide variety of graphical and text-based displays. In addition, it provides the PerfExplorer which uses statistical analysis to detect performance problems and allows to compare multiple application runs.

---

[1] `http://www.in.tum.de/en/forschung/verbundprojekte/clusteraktivitaeten/isar.html`

Vampir [1] provides a trace-based performance analysis framework that converts performance data obtained from a program into different performance views and supports navigation and zooming within these displays. It implements event analysis algorithms and customizable displays which enable interactive rendering of the collected monitoring data. Recently, Vampir's scalability was extended via a parallel analysis server. But still, huge trace files are generated and have to be manually analyzed afterwards.

KOJAK [13] does trace-based analysis of parallel applications. It includes Expert, a component that automatically deduces performance properties from the trace files and provides a user-interface for investigating the found types of properties, the processes where a property occurs, as well as its function or OpenMP region.

SCALASCA [4] (SCalable performance Analysis of LArge SCale Applications) is a scalable trace analysis tool based on KOJAK. It extended KOJAK's approach for automatically searching performance properties by exploiting both distributed memory and parallel processing capabilities. Instead of sequentially analyzing a single global trace file, it uses a technique based on execution replay to gather performance data for detecting formalized properties.

mpiP [7] tool is more oriented towards identifying performance problems for MPI-based applications. It collects the performance data statistically. All information is sampled locally and not exchanged between processes during runtime. The results of the experiments are only merged after program termination as a postmortem approach.

Our approach advances the state of the art in the following aspects:

1. Performs a distributed search of properties.
2. The performance data are processed while the application runs.
3. Periscope knows and exploits the structure of the application. This information is generated by source code instrumentation and is used by the analysis agents.
4. It does not use trace-based analysis and hence avoids large datasets.
5. It has a user-friendly interface that supports remote execution on large and distributed machines and inspection of the application's performance properties.

1.3 Periscope Design

Periscope's design provides mechanisms to search for performance properties in a distributed fashion based on a master agent for management purposes, analysis agents for finding problems on individual processors, and a few other agents responsible for communication. It is devised such that the users can investigate performance problems on large-scale runs delving into the single node especially memory-related performance bottlenecks and MPI/OpenMP issues.

It can be used in interactive application runs as well as in batch runs. The analysis is executed in an iterative fashion, i.e., focusing on more specific and precise properties in multiple executions of a program phase. The program phase is either a special code region marked by the user or the whole code. While in the first case, the execution is suspended when the end of the phase region is reached and the processes are released afterwards for a next execution, in the latter case the application is automatically restarted for another execution of the phase.

In this section, we discuss Periscope's architecture, the agent's data capturing mechanism, different types of search strategies and the Eclipse plugin [3] implementing the user interface.

1.3.1 Periscope Architecture

The Periscope architecture consists of four major entities as shown in Fig. 1.2 that can perform distributed and online-based performance analysis of any scientific application.

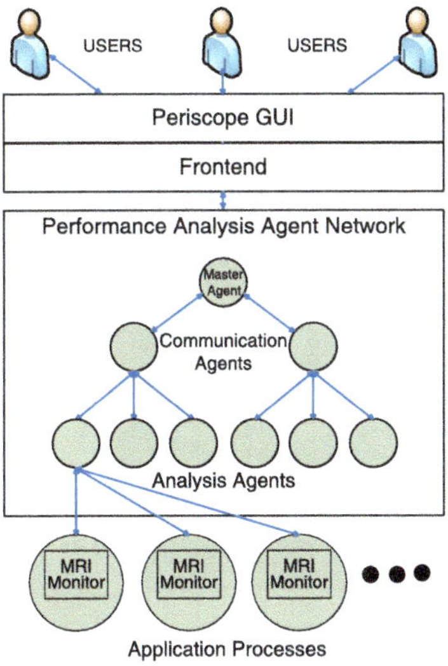

Fig. 1.2: Periscope Architecture

1. The *User-Interface* is a convenient and feature-rich entity that displays the results of the runtime analysis by directly mapping the detected properties to the source code. This entity benefits Periscope users by providing many advanced features like code indexing, refactoring, syntax highlighting, and so forth.
2. The *Frontend* starts the application and analysis agents based on the specifications provided by the user, namely, number of processes and threads and so on. It is responsible for restarting the application whenever the agent network did not complete the search before the application terminated.
3. The *Analysis agent network* consists of three different agents, namely, master agent, communication agent and analysis agent. The *master agent* forwards commands from the frontend to the analysis agents and receives the found performance properties from the individual analysis agents and forwards them to the frontend. The *communication agents* combine similar properties found in their

sibling agents and forward only the combined properties. The *analysis agents* are responsible for performing the automated search for performance properties. During the search they access the monitor linked to the application processes via the Monitoring Request Interface (MRI).
4. The *MRI monitors* provide an application control interface. They deal more with hardware and software sensors to measure the performance data.

1.3.2 Agent's Data Capture Mechanism

The agents play a vital role in the search for performance properties in the processes or threads of the application. The agent design consists of two main parts, namely, the agent and the monitor. The main components of the agents are the agent control, the search strategy, and the experiment control. Figure 1.3 presents the agent design and the sequence of operations involved in capturing performance data. In general,

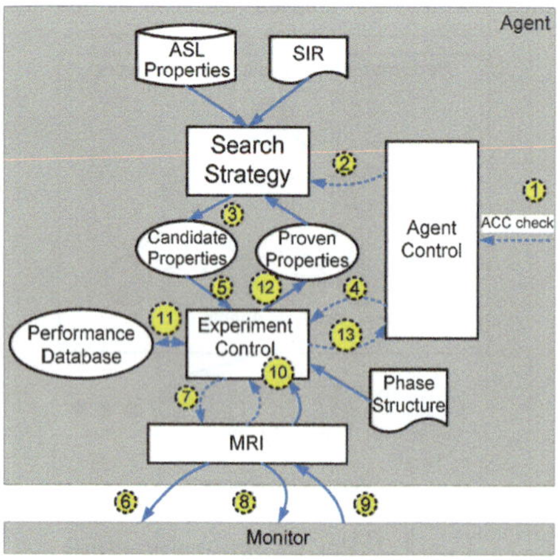

Fig. 1.3: Sequence of operations performed by agents to capture performance data

any scientific application would have computational intensive iterations. The user can mark those regions as user-region or phase-region, so that, it can be used to perform a multistep search. In order to perform such search autonomously, agents are involved in the following phases of the performance data capturing mechanism in Periscope:

1. initialization phase
2. execution phase
3. data collection phase
4. evaluation phase

During the *initialization phase* (1 to 3 in Fig. 1.3), the master agent starts the performance analysis search which triggers the search strategy and initializes the candidate set and triggers a new experiment. The master agent starts the performance analysis via the Agent Control and Command (ACC) message `CHECK`. Before the message is sent, the application is started and suspended in the initialization of the monitoring library linked to the application. In addition, the analysis agent is instructed to attach to the application via the monitor. The initial candidate set is determined by a search strategy based on the set of properties found in the previous step. At the beginning, the set of evaluated properties is empty.

The *execution phase* (4 to 8 in Fig. 1.3) of performance data capture mechanism involves triggering a new experiment, checking for missing performance data, sending MRI requests, releasing the application, and configuring the monitors. After the candidate set is determined, the agent control starts a new experiment. The experiment control accesses all the properties in the candidate set and checks whether the required performance data for proving the property are available. If not, it configures the monitor via new MRI measurement requests. The source code instrumenter used for the insertion of monitoring library calls generates information about the program's regions (main routine, subroutines, loops, etc.) in the Standard Intermediate Program Representation (SIR) format developed by the European American APART working group on automatic performance analysis tools. The requests, such as, measure the number of cache misses in the parallel loop on line 52 in file foo.f, are sent using sockets to the monitoring library. Once all the properties were checked for missing performance data, the experiment is started. The MRI provides an application control interface that allows the experiment control to release the suspended application and to specify when the application is to be suspended again to retrieve the performance data. During program execution, the monitoring library checks whether the end of the current phase is reached and configures hardware and software sensors for measuring the requested performance data.

In the *data collection phase* (9 to 11 in Fig. 1.3), a specific mechanism is used to collect performance data. In this lieu, data are inserted into a summary table. When the application is suspended again the experiment control is informed by the MRI and it retrieves the measured performance data via the MRI into the internal performance database.

Finally, during the *evaluation phase* (12 in Fig. 1.3), the candidate properties are evaluated and the sequence of operations mentioned above are repeated. Evaluation of the candidate performance properties is done by the experiment control and they are inserted into the proven properties set. At the end of this search step, the control is returned to the agent control and the cycle of the analysis steps is repeated.

1.3.3 Search Strategies

Currently, three search strategies are supported: one for analyzing the single-node performance, another for investigating the MPI communication behavior and one combining both types. Characteristic for the first search strategy is that it is a multi-step one while the MPI is a single-step operation.

1.3.3.1 Single-Node performance

There are two strategies for analyzing the code's efficiency based on stall cycle counters. They both use the PAPI library which provides a portable access to hardware performance counters. These strategies give an insight into the pipeline execution and the usage of the cache memory. The first implemented search strategy is the so called *StallCycleAnalysis*. It analyzes the defined phase region for stall cycles and, depending on their number, it can either refine the property hierarchy or stop the analysis. After detecting and proving all performance problems in the current region this strategy continues with the nested code parts and function calls. The newly created hypotheses are based on the detected properties in the parent regions in order to reduce the measurement overhead.

The second available strategy is *StallCycleAnalysisBreadthFirst*. Its fundamental idea is quite similar to the previously discussed stall cycle analysis. The main difference is that this one detects the bottlenecks for all regions in one step. It too requires multiple executions of the phase region to gather all the performance data since only a limited number of hardware counters is available.

1.3.3.2 MPI Communication Behavior

Improving the communication pattern and finding performance imbalance is a crucial optimization step for MPI applications. In order to help the developer in tuning his program, Periscope implements a MPI search strategy. It is able to detect different synchronization problems, such as, late sender and receiver. In contrast to the other strategies, this one gathers all the necessary runtime information in only one run of the phase region. The main reason for this is that the properties have no special hierarchy, thus no further refinement is required.

1.3.3.3 Aggregated Strategy

In addition, a meta strategy called *AllStrategy* exists. It does neither define any initial hypotheses nor provide any refinement algorithms. Its purpose is to combine the strategies for measuring both the single-node performance and the MPI behavior. In the beginning the former group is activated. When the stall cycle analysis is completed, the MPI one is started. The result is a complex profile that gives a deeper insight into the runtime performance of MPI programs.

1.3.4 User Interface

Periscope provides a convenient GUI as shown in Fig. 1.4 aiming at enhancing the analysis and post-processing of the found performance properties. It is developed as a plug-in for Eclipse in order to integrate it with other available programming tools such as the C/C++ Development Environment (CDT), Fortran IDE (Photran), Remote System Explorer (RSE), etc. It currently consists of three interconnected views that present the detected properties and also provide an overview of the instrumented code regions. Due to the textual character of the bottlenecks stored by

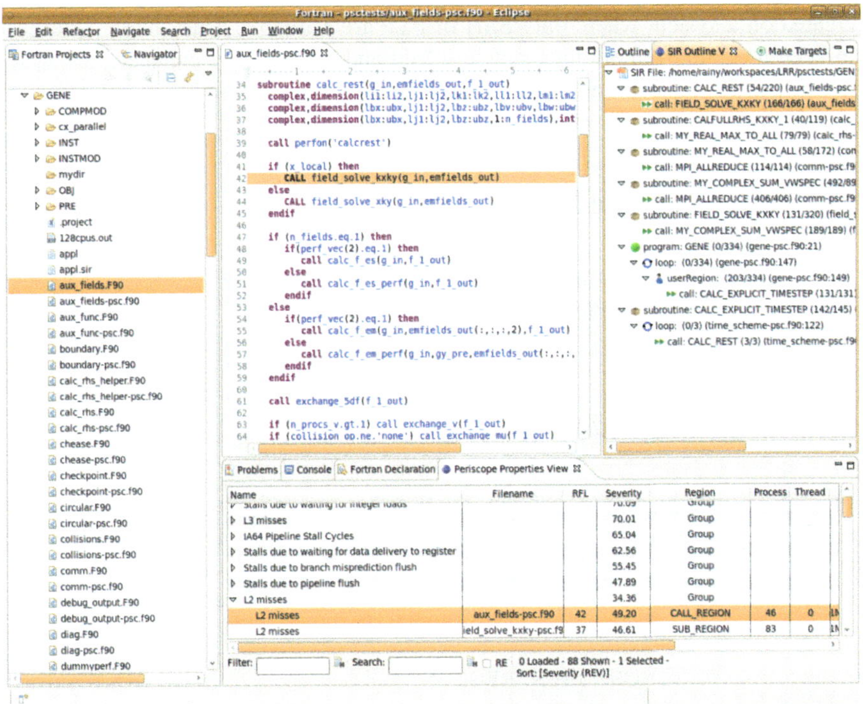

Fig. 1.4: Snapshot of the Periscope GUI

Periscope and their summarized form, a multi-functional table is used for their visualization. To organize the displayed data, so that, maximum knowledge can be gathered out of it, the table provides the following features:

- Multiple criteria sorting algorithm
- Complex categorization utility
- Searching engine using regular expressions
- Filtering operations
- Direct navigation from the bottlenecks to their precise source location using the default IDE editor for that source file type (e.g. CDT/Photran editor).

An outline view for the instrumented code regions that were used in an experiment is also available. Statistical clustering is another key feature of the plug-in that enhances the scalability of the GUI and provides means of conducting peta-scale performance analysis. It can effectively summarize the displayed information and identify a runtime behavior possibly hidden in the large number of properties.

After every performance experiment of Periscope, the tool generates an XML-based file with the detected properties. This file, together with source code of the application and its SIR document, must be organized in a project for the Eclipse IDE so that the GUI can access it and load it for analysis. Because of the integration of the Eclipse File System (EFS) as a file management layer, the most appropriate data storage location can be freely chosen. As a result, a remote project can be created on a system where only SSH or FTP are available. Thus, there is no need to keep local and remote files synchronized and so it greatly enhances the whole tuning process of real-world applications.

1.4 Periscope Usage Model

This section explains the usage scenario of Periscope to find the performance bottlenecks in scientific applications.

1.4.1 Preparing Analysis Run

In order to undergo an analysis run, the user has to configure Periscope by stating the machine and the port number for the internal registry, analysis agents and applications. Optionally, the user can guide the search by stating the phase region via a user region in the code. For example, in Fortran-based applications, the user-region is marked as shown below:

```
!$MON USER REGION
     ...
!$MON END USER REGION
```

It is mandatory that the program has to be instrumented, compiled, and linked with the required libraries (MRIlib and mpiProfilerlib). In the existing application's makefile, the compilation step generating the object files has to be modified such that the compiler is replaced with the command *psc_instrument*. The script will preprocess the files, instrument them, and finally call the compiler for generating the instrumented object files. In addition, the compiler has to be replaced in the link step by *psc_instrument*. Here, it will link also the monitoring library to the executable as well as generate the SIR file with the program's static information.

Instrumentation of subroutines, call sites, user regions, loop regions, OpenMP parallel regions can be switched on/off using the keywords *sub, call, user, loop,* and *par*. This specification can be given for each source file individually via the *psc_config_file* of Periscope.

1.4.2 Starting an Analysis Run

For starting an analysis run, the user has to start a registry service by specifying its alloted port number. On the Altix, the following command is used to start the registry:

```
regsrv.ia64 35000 &
```

The agents and the application processes register with it their location and the ports they use. The Periscope Frontend starts the analysis agent hierarchy. It also runs the application and optimizes the mapping of application processes and agents to processors. The number of processes and threads are specified via *ompnumthreads* and *mpinumprocs* arguments. It will first contact the registry and then start the application. After all application processes registered with the registry, the agent hierarchy starts, the analysis agents connect to the application processes and the search starts. The following command is used for starting an analysis run on our Altix:

```
frontend.ia64 --apprun=~/psctest/GENE/gene
      --mpinumprocs=1024 --strategy=
       StallCycleAnalysis --debug=1
```

The *apprun* parameter specifies the command line to start the application. It will be passed to the mpirun command. Specifying the number of MPI processes is done using the command *mpinumprocs*. The argument *strategy* can be one of the following: MPI, StallCycleAnalysis, StallCycleAnalysisBreadthFirst, and AllStrategy. The command *debug* is for setting the debug level. In addition, the OpenMP threads, SIR filename, application port number, number of processors controlled by a single analysis agent, number of highlevel agents and the timeout can be explicitly specified in the command line.

1.5 Distributed Performance Analysis

As an example for identifying performance bottlenecks with Periscope for a large-scale scientific application, we demonstrate here the performance analysis of a Gyrokinetic Electromagnetic Numerical Experiment (GENE) code. The GENE code [2, 8] is an iterative solver for a non-linear gyrokinetic equations in a 5-dimensional phase space to identify the turbulence in magnetized fusion plasmas. It was developed in the Max Planck Institute for Plasma Physics in Garching. GENE has the capability to run with a large number of processors, such as, 512 or 1024. It consists of 47 source files with 16,258 lines. The code is written in Fortran 95 with MPI-based parallelization.

The experiments were carried out on the ALTIX at LRZ. In the test, the application was executed in eight partitions with 134, 154, 24, 144, 50, 257, 143, 94 processors and one analysis agent in each partition. The GENE code was analyzed on 1, 8, 16, 32, 64, 128, 256, 512 and 1024 processors based on its parallelization needs.

The experimental results revealing the found properties in different code regions of GENE, scalability issues that include single node performance, loading properties, organizing information, and clustering mechanisms are detailed in the following subsections.

1.5.1 Found Properties

The main program, gene.f90 of 47 files is responsible to read the parameters, initialize conditions and calculate the explicit time loops. The user-region indicating the phase for the incremental analysis was marked covering the time step.

The properties found are summarized in Table 1.1. The column headings with 1, 2, 3, 4, and 5 represents the properties, respectively, IA64 pipeline stalls, Stalls due to L1D TLB misses, Stalls due to pipeline flush, Stalls due to waiting for FP registers, and Stalls due to waiting for integer registers. It can be seen that the code suffers more from integer loads compared to floating point registers.

Table 1.1: Found properties in GENE code

Sl.No	File Name	Region	Line Number	Average Severity				
				1	2	3	4	5
1	gene.f90	User Region	149	38.3	6.85	11.1	4.14	17.65
		Call Region	163	37.6	4.41	10.2	3.0	20.4
2	time_scheme.f90	Subroutine	91	37.7	3.1	10.3	2.4	22.2
		Loop Region	122	22.2	*	6.0	2.8	11.1
		Call Region	129	12.4	*	7.7	*	13.2
3	field_s_kxky.f90	Subroutine	37	18.5	*	11.9	*	19.3
		Call Region	65	17.7	*	9.6	*	19.6
3	aux_fields.f90	Subroutine	34	18.3	*	11.5	*	19.8
		Call Region	42	17.9	*	11.9	*	19.4
3	comm.f90	Subroutine	305	17.5	*	11.9	*	19.6
		Call Region	312	17.1	*	1.1	*	*

1.5.2 *Scalability of the Analysis*

Several techniques were introduced in Periscope to support the analysis of large-scale applications.

1.5.2.1 Loading Properties

While analyzing the GENE code with 1024 CPUs, Periscope detected plenty performance properties of 14 different types which were distributed in 4 code regions. In order to enhance the analysis an additional menu entry (see Fig. 1.5a) is provided in the GUI to load only properties with high severity.

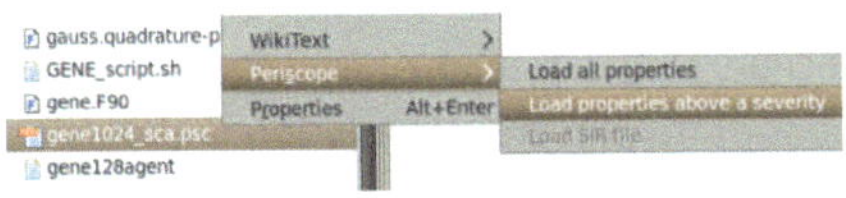

(a) Loading properties above a threshold

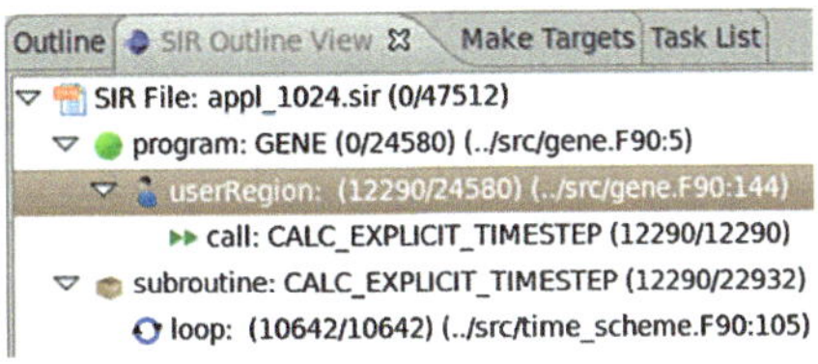

(b) Instrumentation nesting

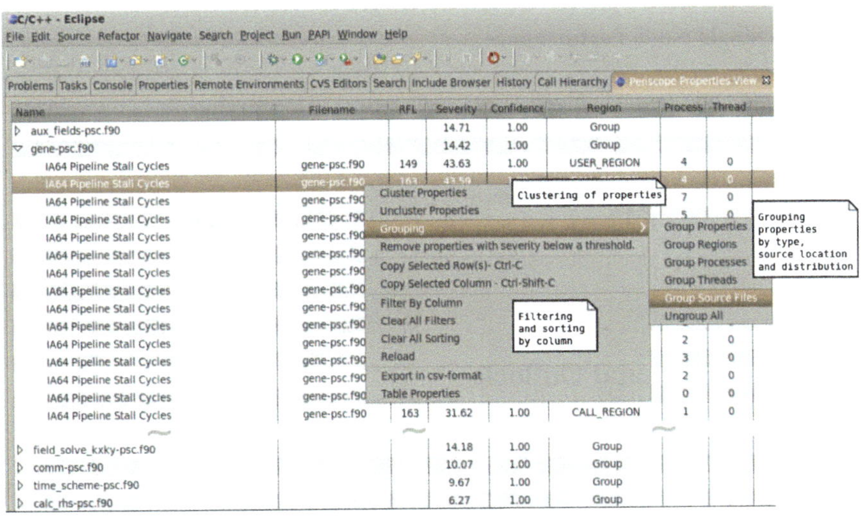

(c) Grouping and clustering the information

Fig. 1.5: Customizing the displayed information

1.5.2.2 Organizing Information

Processing the large amount of properties from a large-scale execution might be a challenging job for the application developers. In order to tackle this problem, Periscope provides a way to group the collected data in different categories according to the type and the location of the performance property as shown in Fig. 1.5c. Furthermore, the displayed data can be filtered from the instrumentation outline view to show only the entries of a selected code region. The view also exposes the complete instrumentation nesting (see Fig. 1.5b).

1.5.2.3 Clustering Mechanism

The GUI for Periscope provides a basic multivariate statistical clustering support that is based on a data-mining workbench called Weka. This feature is activated from the context menu as shown in Fig. 1.5c. Currently, we used the *SimpleKMeans* algorithm to group the detected performance bottlenecks based on their distribution on the CPUs and their respective code regions. During the analysis of the GENE code with 1024 processors, the algorithm generated a few clusters of similarly behaving processors. It scaled down the amount of displayed information and so made it easier to uncover runtime behavior that was previously hidden in the huge dataset.

1.5.2.4 Single Node Performance

The single node performance for large-scale run (Fig. 1.6) illustrates the efficiency of the computation. For the run of GENE with 1024 processors, Periscope identified that processor 112 had an unexpected peak of 13 severity points from its average for IA64 pipeline stalls, and processor 868 had 19 severity points for stalls due to L1D TLB misses. The figure shows that the processors 834 to 862 had comparatively good performance due to less severity than other processors.

1.6 Conclusion and Outlook

Periscope is an online-based performance analysis tool for detecting performance problems in large-scale scientific applications. The search is executed in an incremental fashion by either exploiting the repetitive behavior of program phases or by restarting the application several times. The GUI provides an excellent user-friendly interface for the developers.

This article presented the agent design to find the wide range of bottlenecks, Periscope's GUI with excellent features, such as, remote execution, grouping large datasets for scalability issues, and clustering of found properties. In addition, the tool was demonstrated with the real-world large-scale scientific application, namely,

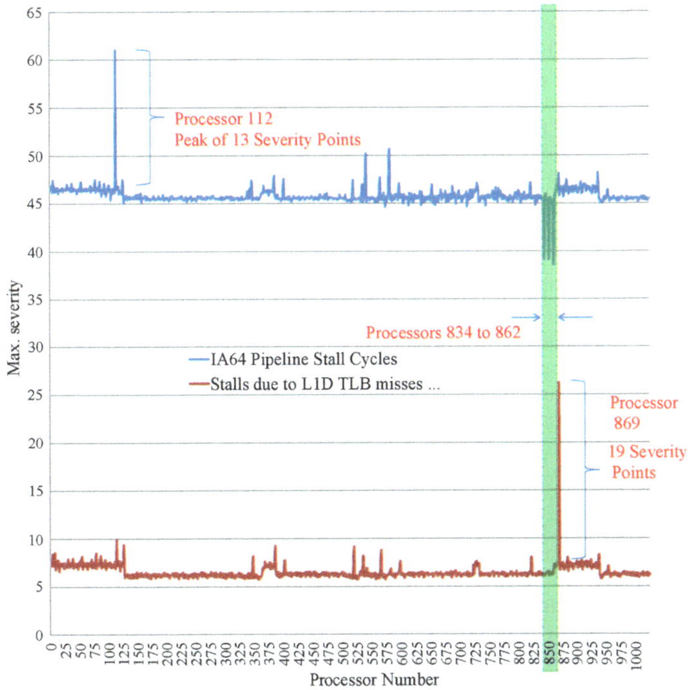

Fig. 1.6: Single node performance of GENE when running in Altix for 1024 processors

GENE to find the single node performance on large-scale runs, its scalability issues and the found properties. Since the analysis was carried out with the Periscope GUI, the tool was found to be user-friendly and efficient for displaying the necessary information.

While the largest test runs with Gene were executed for 1024 processes due to the availability of test data for that size, we did other experiments with up to 4000 processors on the Altix. These experiments helped us in improving the scalability of Periscope considerably.

Periscope is currently running on Itanium Altix, Power 6, and Linux X86-based parallel machines. In the near future, we plan to port Periscope to BlueGene/P.

References

1. Andreas Knüpfer, Holger Brunst, Jens Doleschal, Matthias Jurenz, Matthias Lieber, Holger Mickler, Matthias S. Müller and Wolfgang E. Nagel. The Vampir Performance Analysis Tool-Set. In *Proc. of the 2nd Int. Work. on Parallel Tools for HPC, HLRS, Stuttgart*, pages 139-155, Springer Publications, July 2008.

2. Chen, Y. and Parker, S. E. A δf particle method for gyrokinetic simulations with kinetic electrons and electromagnetic perturbations. In *Comput. Phys. 189, 2 (Aug. 2003)*, DOI: `http://dx.doi.org/10.1016/S0021-9991(03)00228-6`. pages 463-475, 2003.
3. Eric Clayberg and Dan Rubel. Eclipse Plug-ins. In *Addison-Wesley Professional, ISBN 978-0-321-55346-1* pages 107-135, 2008.
4. Markus Geimer, Felix Wolf, Brian J. N. Wylie, and Bernd Mohr. Scalable parallel trace-based performance analysis. In *Proc. of the 13th Eur. PVM/MPI Group Meeting on Recent Advances in Parallel Virtual Machine and Message Passing Interface (EuroPVM/MPI 2006), pages 303–312, Bonn, Germany, 2006.*
5. M. Gerndt and K. Fürlinger. Specification and detection of performance problems with ASL. *Conc. and Computation: Prac. & Exp.*, 19(11):1451–1464, Aug 2007.
6. Michael Gerndt and Edmond Kereku. Search strategies for automatic performance analysis tools. In Anne-Marie Kermarrec, Luc Boug, and Thierry Priol, editors, *Euro-Par 2007*, volume 4641 of *LNCS*, pages 129–138. Springer, 2007.
7. Jeffrey Vetter and Chris Chambreau. mpiP: Lightweight, Scalable MPI Profiling. `http://mpip.sourceforge.net`, 2008.
8. F. Jenko. Massively parallel vlasov simulation of electromagnetic drift-wave turbulence. In *Comp. Phys. Comm. 125* 2000.
9. B.P. Miller, M.D. Callaghan, J.M. Cargille, J.K. Hollingsworth, R.B. Irvin, K.L. Karavanic, K. Kunchithapadam, and T. Newhall. The Paradyn parallel performance measurement tool. *IEEE Computer, Vol. 28, No. 11, pp. 37-46*, 1995.
10. Philip C. Roth and Barton P. Miller. The distributed performance consultant and the sub-graph folding algorithm: On-line automated performance diagnosis on thousands of processes. In *Proceedings of the ACM SIGPLAN Symposium on Principles and Practice of Parallel Programming (PPoPP'06)*, March 2006.
11. Shajulin Benedict, Matthias Brehm, Michael Gerndt, Carla Guillen, Wolfram Hesse and Ventsislav Petkov. Automatic Performance Analysis of Large Scale Simulations. In *PROPER 2009, (in press), Springer Publishers* 2009.
12. Sameer S. Shende and Allen D. Malony. The TAU parallel performance system. *International Journal of High Performance Computing Applications, ACTS Collection Special Issue*, 2005.
13. Felix Wolf and Bernd Mohr. Automatic performance analysis of hybrid MPI/OpenMP applications. In *Proceedings of the 11th Euromicro Conference on Parallel, Distributed and Network-Based Processing (PDP 2003)*, pages 13–22. IEEE Computer Society, February 2003.

Chapter 2
Comprehensive Performance Tracking with Vampir 7

Holger Brunst, Daniel Hackenberg, Guido Juckeland, and Heide Rohling

Abstract Vampir 7 is a performance visualization tool that provides a comprehensive view on the runtime behavior of parallel programs. It is a new member of the Vampir tool family. This new generation of performance visualizer combines state-of-the-art parallel data processing techniques with an all-new graphical user interface experience. This includes fast local and remote event data browsing, searching, filtering, clustering, and summarization. The software is ported to Unix, Windows, and Apple platforms. This article gives an overview of the novel techniques and features of Vampir 7.

2.1 Introduction

Performance tracking is about understanding and improving the inner workings of software for complex computer infrastructure. The presented *software microscope* Vampir addresses the visualization of concurrent software processes at user definable levels of detail. The main motivation for our activities is scientific curiosity, efficient usage of in-house compute resources, and satisfied customers at our compute center. Certainly, we also support the common argumentation line that stresses the strategic role of Supercomputing and HPC, complex system architectures, painful programming techniques, and the resulting need for supportive tools. Then again, it is probably best to discuss this subject with HPC center managers and parallel application developers rather than tool developers like we are.

Holger Brunst, Daniel Hackenberg, Guido Juckeland, Heide Rohling
Centre for Information Services and High Performance Computing
Technische Universität Dresden
01062 Dresden, Germany
e-mail: {`holger.brunst, daniel.hackenberg, guido.juckeland, heide.rohling`}`@tu-dresden.de`

M.S. Müller et al. (eds.), *Tools for High Performance Computing 2009*,
DOI 10.1007/978-3-642-11261-4_2, © Springer-Verlag Berlin Heidelberg 2010

This article gives an overview of the new Vampir 7 performance data browser and summarizes accomplishments of the Vampir team during the past two years, focusing on the enhanced visualization of time-dependent behavior of accelerated hybrid applications and new performance data sources like the energy consumption of system components.

2.2 Overview

The Vampir 7 performance visualization tool combines modern event processing techniques with a fully redesigned graphical user interface. It is portable and available for many HPC systems. Previously available only for Unix-based systems, Vampir 7 is a complete re-design that is based on a 15 years-plus product history [9].

Vampir provides an easy-to-use analysis framework (see Fig. 2.1), which enables developers to quickly display sequential and parallel program behavior, at customizable levels of detail. This helps developers to analyze their programs, find and identify performance problems, and supports them in producing optimized, more efficient applications:

- It converts performance data obtained from a program into different performance views.
- It supports navigation and zooming within these displays.
- It helps to identify inefficient parts of code.
- It leads to more efficient programs.

Vampir's multi-document GUI offers support for common performance chart types, including timeline and profile. Timeline charts allow studying, debugging, and tuning of the control flow of a parallel application. Adaptive profile charts enable load balancing and subroutine optimization. The communication performance can be analyzed for either individual communication partners or for an entire communication network. In addition to performance tuning, the tool also helps to analyze consistency problems, including communication mismatches, deadlocks, race conditions, and false sharing.

2.3 Recent Developments in the Graphical User Interface

2.3.1 Custom Side by Side Chart Arrangement

The value of individual performance charts can be increased by connecting and correlating them with other charts. Vampir 7 supports a mode of operation, which allows to display multiple time correlated charts side by side. Charts that display a

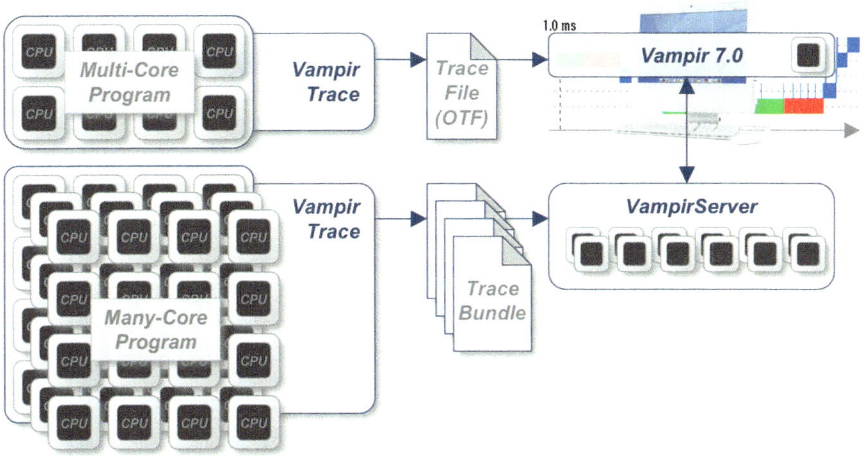

Fig. 2.1: Vampir 7 framework overview. VampirTrace (left) records relevant actions of a given parallel program. The resulting trace file is translated into graphical charts with the Vampir GUI (top right). Extensive program runs on many cores can result in potentially large trace data bundles (collection of files). They can be processed with the VampirServer middle-ware (bottom right).

timed sequence of events such as the "Master Timeline" or the "Process Timeline" are always aligned vertically. This alignment ensures that the temporal relationship of events is preserved across chart boundaries.

The user can arrange the placement of the charts according to his preferences by dragging them into the desired position. When the left mouse button is pressed while the mouse pointer is located above a chart title, the layout engine highlights potential positions (chart edges) with a red bar. As soon as the user releases the left mouse button, the chart arrangement will be changed according to his intentions. The entire procedure is depicted in Fig. 2.2.

The dynamic chart layout engine allows to increase or decrease the space that is used by individual charts. Charts of particular interest may get more space in order to render information in more detail. The main document window can host an arbitrary number of charts. Charts can be added by clicking on the respective "Charts" toolbar icon. With a few more clicks, charts can be combined to a custom chart arrangement as depicted in Fig. 2.3. Customized layouts are stored per trace data set in order to allow the easy continuation of interrupted analysis sessions and presentations.

Every chart can be undocked or closed by clicking the dedicated icon in its upper right corner as shown in Fig. 2.3. Undocking a chart detaches the chart from the current arrangement and presents it in its own separate window.

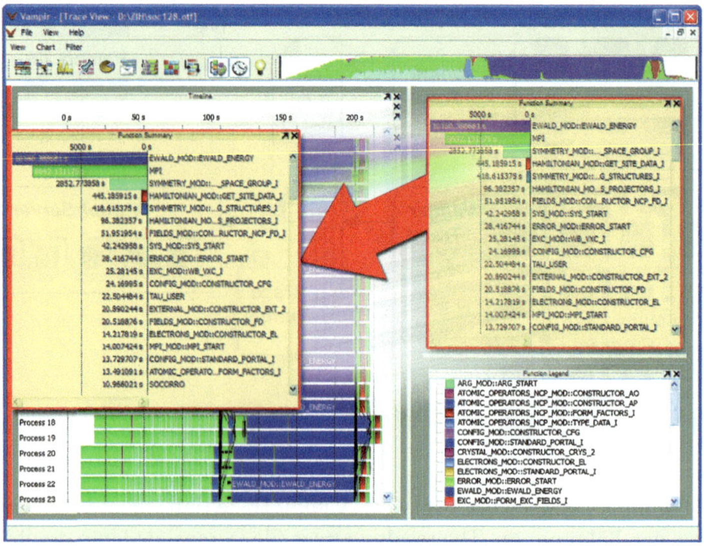

Fig. 2.2: Moving and arranging charts in Vampir 7. All available performance charts can be dragged to arbitrary positions. The layout engine avoids chart overlapping. Furthermore, all charts depict the same time interval, process, selection, and color encoding. Automatic horizontal alignment is used across multiple timeline charts.

2.3.2 Counter Data Timeline

Counter data are collected over time. Typically events like floating point operations or cache misses are counted. Counter values can be used to store not just hardware performance counters but arbitrary sample values. There can be counters for different statistical information as well, for instance counting the number of function calls or a value in an iterative approximation of the final result. Counters are defined prior to the instrumentation of the application and can be individually assigned to processes.

An example timeline chart with counter data is shown in Fig. 2.4. The chart shows performance rates that were derived from the counter "floating point operations" for a given time interval. The graph shows the floating point rates of the respective program phases. The measurements were carried out at a very high precision. Due to the limited display resolution, two individual graphs characterize the performance over time. The red graph identifies the maximum performance rates that were measured over time whereas the average performance rate is depicted in grey below the upper graph. Optionally, a minimum graph (not depicted) can be added to the chart showing the poorest achieved performance rates.

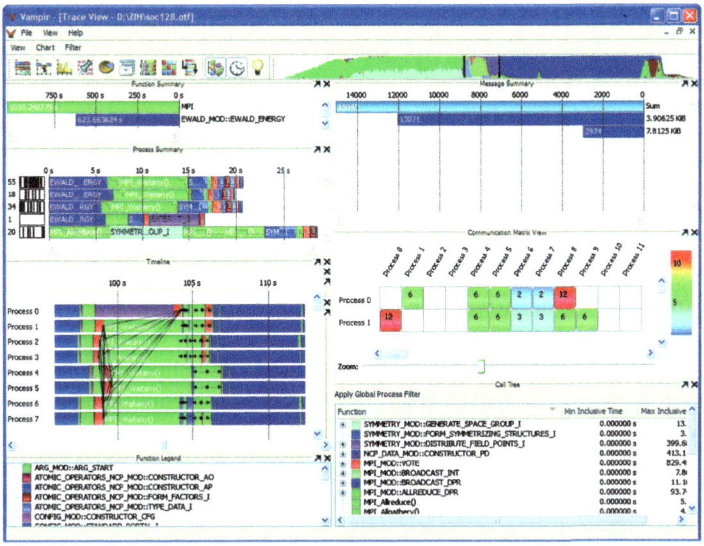

Fig. 2.3: A custom chart arrangement in the main document window of Vampir.

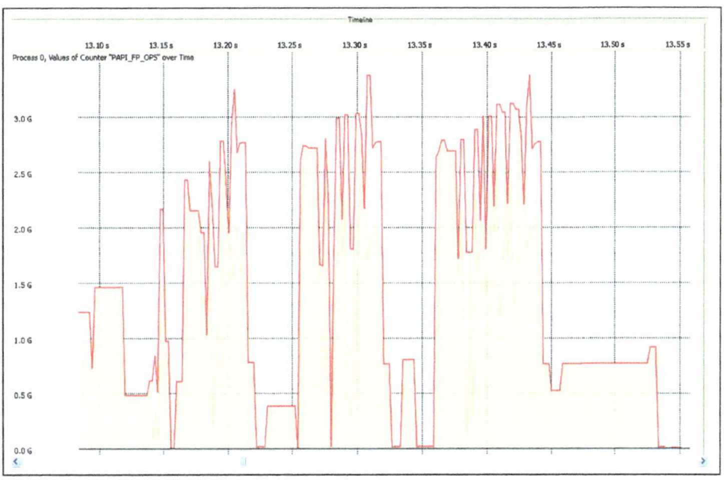

Fig. 2.4: Counter Data Timeline. The graph shows the floating point rates of the respective program phases. The maximum performance rates are represented by the red graph whereas the average performance is depicted as a grey line below.

2.3.3 *Performance Markers*

A new feature in Vampir 7 is the visualization of custom *performance markers*. Markers can be used to highlight arbitrary hot spots in trace files. They were designed to improve the integration with automized performance analysis tools like Scalasca [2] or Periscope [11]. For critical points marker events can be set, which consist of a timestamp, a textual description, and the process or process group. Marker events have a specific type, e.g. "Error", which is defined by the user. Markers are not interpreted by Vampir but displayed with all their properties in a display called "Marker View" and drawn as little triangles in the timeline views. The marker view provides a sorted list of all marker events. The process, time and description of all events are shown as depicted in Fig. 2.5. They are sorted by their groups and markers. Because of the tree structure the user easily can navigate through it. Vampir supports "Error", "Warning", and "Hint" as groups. The rest is classified as "Unknown".

The triangles in the Master Timeline and Process Timeline are colored according to the group they belong to. The changeable color for "Errors" is by default red and for "Warnings" orange, to illustrate the severity. By clicking on a marker event in the "Marker View" the corresponding events in the timelines are highlighted and vice versa. In addition to this, it is possible to highlight all events of the same marker or group.

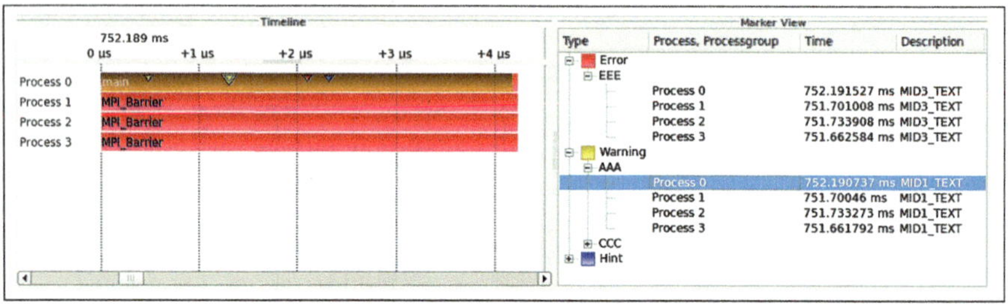

Fig. 2.5: Performance markers highlight hot spots during the execution of a program. Individual markers can be selected in a hierarchical list (right). Their occurrence in time is depicted as colored triangles in the respective timeline views (left).

2.3.4 Clustering of Performance Data

Traditionally, performance profiles list the following information for N_f functions f of a program: accumulated time, number of invocations, and average time. In case of a parallel program, a separate profile is provided for each concurrent task τ (i. e., process or thread) of the overall N_τ tasks. The time parameter is typically available as an inclusive (including the time spent in subroutines) or exclusive (excluding the time spent in subroutines) value. Assuming a highly parallel code with $N_\tau > 1{,}000$, only a few monitored functions are sufficient to produce very complex results. The situation becomes worse for more sophisticated profiling techniques like call graph

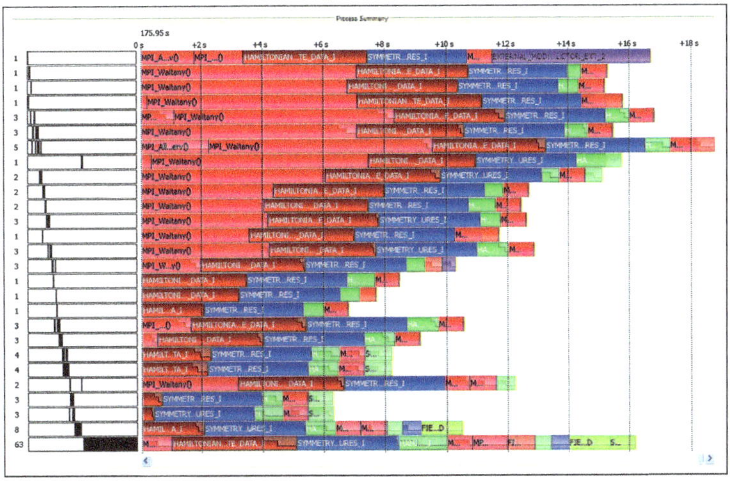

Fig. 2.6: A clustered parallel profile of originally 128 concurrent tasks. Similar profiles are merged to 27 representatives showing average values. The profile legend on the left shows a graphical task distribution vector ranging from 1 to 128. The number in front of each vector identifies the number of clustered profiles represented by the respective bar.

or call path profiling [5, 7]. These approaches collect data depending on the call site (or even the entire call path) of a function call resulting in many more measured values per function. Multiplying the number of results per task by the overall number of tasks quickly leads to a complexity that becomes unmanageable. The thorough analysis of profile data from many highly parallel applications reveals that most concurrent tasks in an application follow the same execution pattern. This is mainly due to symmetric algorithms and data dependencies leading to global or partial task synchronization. Resulting redundancies in profile data can be used to limit the data visualization to representative tasks which reflect a certain class/pattern of behavior. The idea is to cluster similar task profiles. Objects inside such a cluster are very similar while objects in different clusters should be very different.

Henceforth, task profiles are represented by vectors $\mathbf{x} = (x_1, \ldots, x_{N_f})$ where x_i is the numerical result of an arbitrary experiment, e. g., the average duration of a specific function f. To evaluate whether a set of profiles is similar enough to be considered a cluster, a distance measure $d(\mathbf{x}, \mathbf{y})$ is needed between two profiles $\mathbf{x}$ and $\mathbf{y}$. Assuming a k-dimensional Euclidean space, the distance between two objects, $\mathbf{x} = (x_1, x_2, \ldots\ldots, x_{N_f})$ and $\mathbf{y} = (y_1, y_2, \ldots\ldots, y_{N_f})$ may be defined using the Euclidean distance:

$$d(\mathbf{x}, \mathbf{y}) = |\mathbf{x} - \mathbf{y}| = \sqrt{\sum_{i=1}^{N_f} (x_i - y_i)^2} \tag{2.1}$$

Given N_τ the number of profiles and N_k the number of clusters to create, one reasonable approach among others (hierarchical, density-based, and grid-based) is to use a partitioning approach. Based on an initial partitioning, an iterative relocation technique attempts to improve the partitioning by moving objects from one group to another. To achieve global optimality, partition based clustering requires the exhaustive enumeration of all of the possible partitions. A popular heuristic called the *k-means algorithm* is used to obtain suboptimal results much faster. The k-means algorithm produces a set of N_k clusters that minimizes the squared-error criterion. The number of clusters is configurable by the user.

Figure 2.6 depicts a clustered function profile. The number of clusters has been set to $k = 27$ which was the maximum number of bars that fitted in given chart height. Each bar represents a class of similar profiles. Another major difference to traditional profile bar charts is the representation of task identity and distribution. Due to the nature of the clustered profiles, a simple numbering scheme is no longer applicable to the profile bars. Therefore, each profile bar is now preceded by a task distribution vector ranging from Task 1 to Task N_τ.

2.4 New Performance Data Sources

2.4.1 Accelerators

Multi-core technology ultimately found its way into main-stream processor families. Today, the growing popularity of hardware accelerated computing even further increases the complexity of the software development process. Especially the correct and efficient usage of the increasingly complex memory hierarchy is mandatory to obtain best application performance. With traditional profilers, the impact of code tuning activities can often only be observed from a rather external point of view. The exact identification and location of the culprit is often impossible and the real reasons behind a performance deficiency remain obscure.

Accelerators have very specific hardware and software characteristics. Yet, they do have the following common optimization challenges from a programmer's perspective:

1. Memory utilization (Data transfers, global/host memory access)
2. Core/Unit utilization
3. Task/Thread synchronization

While there exist several major architectural differences, Table 2.1 identifies several similarities between accelerators and traditional multi-core CPUs [6, 10]. We introduce common terms like host processor, local memory, or synchronization for common architectural features to simplify further discussions.

Table 2.1: Comparison of accelerator features for the IBM Cell, the NVIDIA G200b GPU and a general purpose multi-core CPU

Feature	IBM Cell	NVIDIA GPU	Multi-core CPU
Host processor	PPE	CPU	one core
Accelerator	SPE	GPU	all cores
Local memory	256 KB local store	16 KB shared memory	local cache (L1/L2)
Device memory	–	on graphics board	shared Cache (L3)
Data transfer	DMA transfers	CUDA memcopy	shared address space
Synchronization	mailbox messages	CUDA thread synchronization	shared address space
Accel. program	SPE program	CUDA kernel	threaded program

2.4.1.1 Monitoring

Despite the similarities with standard multi-core processors, acquisition, processing, presentation, and interpretation of performance data is not an easy task in many respects. Existing tools need to be extended to deal with the new hardware and data complexity if a realistic assessment is intended. Our combined monitoring and visualization approach makes the next step in tool evolution towards a highly improved level of detail, precision, and completeness. Vampir's performance monitor [8] now also records specific events of two accelerator APIs, namely IBM's Cell and NVIDIA's CUDA.

First of all, the thread creation on the respective cores needs to be detected and propagated to the host processor which acts as a control unit and supervisor. This includes the synchronization of the individual hardware timers across multiple processors and their cores. During program execution, data transfers, remote access, and user functions (kernels) are logged by customized wrapper libraries. Accelerated program sequences are logged by customized accelerator monitors. The fact that host processor and accelerator use disjoint memory regions is an additional challenge. Thus, log entries that are generated on the accelerator have to be transferred to the host memory and merged into the context of the execution log of the whole application.

The limited amount of local memory on Cell's SPEs requires sophisticated techniques such as double buffering and post-mortem log generation to keep program perturbation low [3]. GPUs, on the other hand, do not offer logging interfaces for local memory accesses. While they offer large amounts of device memory to capture time stamps for occurring events, the available performance critical events are currently limited to start- and end times of CUDA kernels and non-coalesced device memory accesses. This is due to the fact that more detailed performance hooks into CUDA are not yet publicly available.

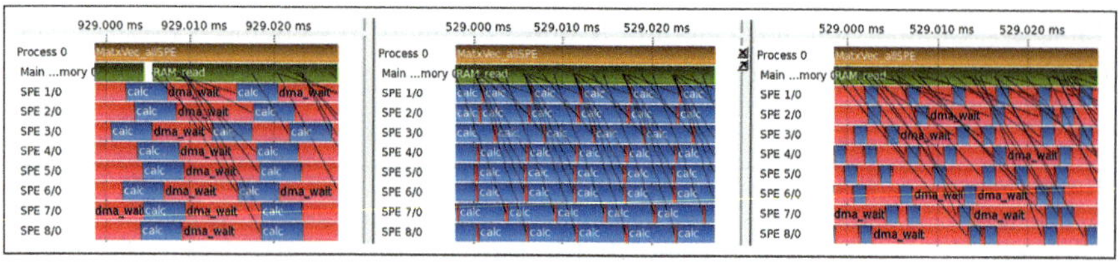

Fig. 2.7: Matrix vector multiplication on Cell: no optimization (left), after memory optimization (center), and after computational optimization (right)

2.4.1.2 Visualization

The program flow and performance graphics in this article were generated by the new Vampir 7 GUI. For a better understanding, some aspects of the accelerator architectures have to be explained with respect to the applied visualization. The host process and its accelerated parts are individually assigned to horizontal bars that change in color to reflect different program regions (e. g. function calls). An important point is to correctly display the target of a data transfer.

Figure 2.7 shows a parallel sparse matrix-vector multiplication on Cell. Calculations (`calc`) and stalls due to DMA transfers (`dma_wait`) can be easily identified. The unoptimized case (Fig. 2.7 left) even shows a phase where not a single accelerator core accesses main memory. We applied two optimizations that overlap DMA transfers with computation (double buffering) and improve the memory bandwidth (128 Byte buffer alignment). The result in Fig. 2.7 (center) shows no `dma_wait` phases anymore. Now, the program is obviously computationally bound. In a next step, a computational optimization is likely to improve the overall performance. We chose loop unrolling as it typically has a significant effect on SPE programs. After this optimization the algorithm is memory bound again as indicated by the numerous `dma_wait` phases in Fig. 2.7 (right). The visualization helps to quickly understand that another computational optimization would have no effect. Further improvements would require to optimize or minimize the memory access.

The second example is a parallel Particle-in-Cell (PIC) code [1] that was ported to NVIDIA GPUs using the CUDA API. Figure 2.8 shows the interaction of three GPUs and their corresponding host CPUs, which in turn also use MPI and pthreads to communicate and distribute work.

Processes 1, 2 and 3 refer to the main program thread that transfers data between the GPU and the main memory. It is dominated by `CUDA_MEMCPY_SYNC` calls because the main program waits for the execution of the GPU kernels when calling a blocking operation such as `CUDA_MEMCPY_SYNC`. Threads 1:2, 2:3 and 3:2 are responsible for visual data output of the GPU memory while threads 1:3, 2:2 and 3:3 are solely responsible for inter-node communication via MPI. The execution state of the CUDA kernels is visualized by the green bars named CUDA[0] 1:1, CUDA[0] 2:1 and CUDA[1] 3:1 where the number in brackets identifies the corresponding node.

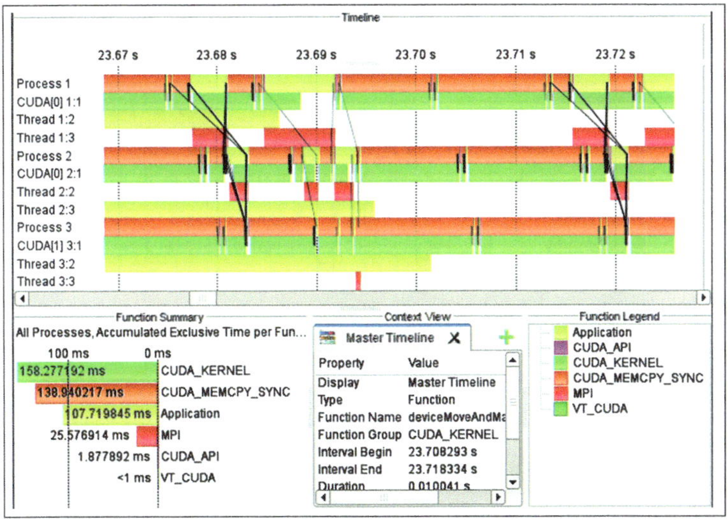

Fig. 2.8: Event log visualization of a 1024×3072 cells Particle-in-Cell (PIC) simulation run with three GPUs on two cluster nodes

The timeline display reveals that the execution of the CUDA kernels (and therefore the PIC computation on the GPUs) is barely interrupted by communication calls and the computational load on the GPUs is well balanced as a result of our performance optimization efforts.

2.4.2 *Energy Consumption*

The energy efficiency of computing resources is among the main challenges of processor and system designers. With the advent of dynamic voltage and frequency scaling (DVS/DFS), a system's energy consumption may vary significantly during application runs. In this context, important factors of an application's runtime are:

- idle phases of (a subset of all) processes that cause DVS/DFS
- usage of on-chip resources like different cache levels or ALUs/FPUs
- usage of off-chip resources like main memory or I/O

Recent CPU designs even extent traditional DVS/DFS as they automatically overclock CPU cores with very high load ("Turbo Boost") although this technique is known to be disadvantageous in terms of energy efficiency.

In order to allow a detailed analysis of these effects, we extended the Vampir tool set to include power measurement data into event logs and to visualize this data. In order to keep additional requirements to the performance browser at a minimum, we use standard Vampir event counters to store the power consumption information.

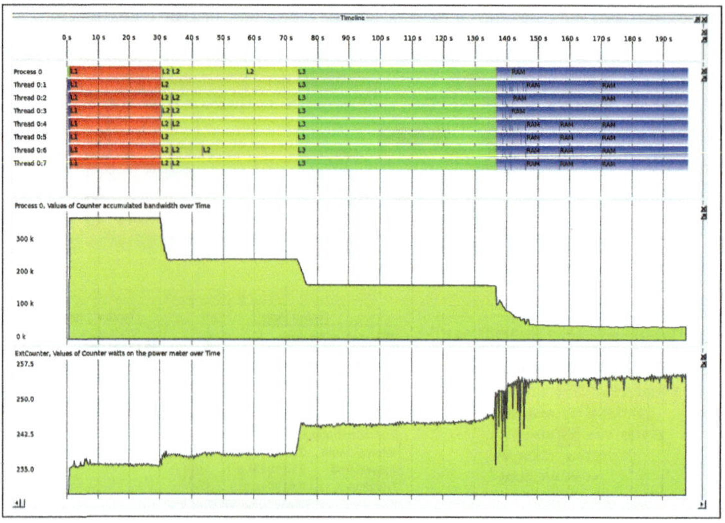

Fig. 2.9: Energy consumption of a multi-core system during the execution of a memory bandwidth benchmark. From top to bottom: a) timeline chart of 8 threads accessing L1, L2, L3, and main memory, b) total bandwidth over time, and c) power consumption of the system over time.

However, our performance monitor Vampir had to be extended in order to record this data during application runs and include it in the event logs. A newly developed infrastructure for external counters is in place to retrieve and store measurement data from power meters and provide it upon request to Vampir's application monitor.

We have tested our approach with several benchmark runs on a dual socket Intel Xeon 5570 (Nehalem-EP, quadcore) node. We measured the power consumption of this node using a ZES LMG 95 power meter with a 20 Hz sampling rate. In this test we focus on evaluating the energy requirements when accessing different levels of the memory subsystem namely the L1, L2, and L3 caches and the main memory. The target application is therefore a set of sophisticated x86_64 memory benchmarks that allows to access data in specific locations [4]. Figure 2.9 shows the visualization of such an event log. It should be noted that the pthreads-parallel benchmark uses all 8 cores to perform identical tasks. The four different memory levels are clearly visible in the timeline view as well as in the two counter data timelines below. The upper graph depicts the memory bandwidth that was achieved when accessing a certain cache level or main memory. The second graph illustrates that the power consumption of the system varies depending on which memory resources are used by the application. Likewise, we can analyze the power consumption of larger MPI-parallel systems.

2.5 Summary and Outlook

This article describes the recent developments of the new Vampir 7 performance analysis framework. It focuses on the new design of the graphical user interface, which introduces many novelties relevant for performance browsers in general. Furthermore, the tapping and processing of new performance data sources, namely hardware accelerators and energy meters, has been described, which is likewise relevant to the HPC community. In the near future, we expect to further improve the support for GPU based accelerators through vendor performance APIs. Furthermore, an extended scripting interface based on DBUS [12] will be incorporated into the performance visualizer. A customizable event search engine and the comparison of successive trace runs are active research topics.

References

1. Birdsall, C.K., Langdon, A.B.: Plasma Physics via Computer Simulation (Series in Plasma Physics). Taylor & Francis, 1 edn. (October 2004)
2. Geimer, M., Wolf, F., Wylie, B., Mohr, B.: Scalable parallel trace-based performance analysis. Recent Advances in Parallel Virtual Machine and Message Passing Interface (2006)
3. Hackenberg, D., Brunst, H., Nagel, W.E.: Event tracing and visualization for Cell Broadband Engine systems. In: Euro-Par (2008)
4. Hackenberg, D., Molka, D., Nagel, W.E.: Comparing cache architectures and coherency protocols on x86-64 multicore SMP systems. The 42nd Annual IEEE/ACM International Symposium on Microarchitecture (October 2009)
5. Hall, R.J.: Call path profiling. In: Proc. Fourteenth International Conf. on Software Engineering. IEEE Computer Society (1992), ftp://ftp.research.att.com/dist/hall/papers/cpprof/icse92.ps
6. IBM: Software Development Kit for Multicore Acceleration Version 3.0: Programmer's Guide (2007)
7. Malony, A.D., Shende, S., Morris, A.: Phase-based parallel performance profiling. In: PARCO. pp. 203–210 (2005)
8. Müller, M.S., Knüpfer, A., Jurenz, M., Lieber, M., Brunst, H., Mix, H., Nagel, W.E.: Developing scalable applications with Vampir, VampirServer and VampirTrace. In: Parallel Computing: Architectures, Algorithms and Applications (2007)
9. Nagel, W.E., Arnold, A., Weber, M., Hoppe, H.C., Solchenbach, K.: VAMPIR: Visualization and analysis of MPI resources. Supercomputer 12(1), 69–80 (1996), http://www.zam.kfa-juelich.de/zam/docs/printable/ib/ib-95/ib-9528.ps
10. NVIDIA: CUDA Compute Unified Device Architecture - Programming Guide (2007)
11. Technical University of Munich: Periscope, http://www.lrr.in.tum.de/~gerndt/home/Research/PERISCOPE/Periscope.htm
12. William, T., Mix, H., Mohr, B., Voigtländer, F., Menzel, R.: Enhanced performance analysis of multi-core applications with an integrated tool-chain. In: Parallel Computing 2009 (PARCO). Lyon (2009), to be published

Chapter 3
Performance Analysis and Workload Characterization with IPM

Karl Fürlinger, Nicholas J. Wright, and David Skinner

Abstract IPM is a profiling and workload characterization tool for MPI applications. IPM achieves its goal of minimizing the monitoring overhead by recording performance data in a fixed-size hashtable resident in memory and by carefully optimizing time-critical operations. At the same time, IPM offers very detailed and user-centric performance metrics. IPM's performance data is delivered as an XML file that can subsequently be used to generate a detailed profiling report in HTML format, avoiding the need for custom GUI applications. Pairwise communication volume and communication topology between processes, communication time breakdown across ranks, MPI operation timings, and MPI message sizes (buffer lengths) are some of IPM's most widely used metrics. IPM is free and distributed under the LGPL license.

3.1 Introduction

Performance analysis and workload characterization serve individual developers and computing centers to understand the performance characteristics of applications. Some important goals of this process are the identification and elimination of performance bottlenecks as well as the development of an understanding of how

Karl Fürlinger
Computer Science Division, EECS Department
University of California at Berkeley
Soda Hall 515
Berkeley, California 94720, USA
e-mail: fuerling@eecs.berkeley.edu

Nicholas J. Wright, David Skinner
NERSC Center
Lawrence Berkeley National Laboratory
Berkeley, California 94720, USA
e-mail: {deskinner, njwright}@lbl.gov

M.S. Müller et al. (eds.), *Tools for High Performance Computing 2009*,
DOI 10.1007/978-3-642-11261-4_3, © Springer-Verlag Berlin Heidelberg 2010

an application scales as the number of processing elements is increased. This paper describes the Integrated Performance Monitor (IPM), an MPI profiling and workload characterization tool that has very low overhead yet is able to deliver important user-centered metrics in detail.

The rest of this paper is organized as follows: In Sect. 3.2 we provide an overview of IPM's hashtable based monitoring approach. In Sect. 3.3 we describe how IPM is used to monitor an application and which kinds of performance data are delivered by IPM. In Sect. 3.4 we study the scalability of MILC (a quantum-chromodynamics code) with IPM. We discuss related work in Sect. 3.5 and conclude with an outlook on planned features and enhancements for IPM in Sect. 3.6.

3.2 Overview

We assume the following general model of an MPI application for the purpose of performance monitoring: The application is composed of n *processes*, each identified by an integer in $[0,\ldots,n-1]$, its *rank*. A set of events $E_i \subseteq E$ happen in each process i. We do not further formally specify what the events are, but we assume they occur at a certain time and have duration. Each event e has an associated *signature* $\sigma(e) \in S$ which captures the characteristics we are interested in. $\sigma : E \mapsto S$ is the signature function. Concretely we think of a signature $\sigma(e)$ as a k-tuple $\sigma(e) = (\sigma^1(e), \sigma^2(e), \ldots, \sigma^k(e))$, where each $\sigma^j()$ is a signature *component*. Useful components of signature functions are listed in Fig. 3.1.

Signature Component	Function	Data type	Typical Size (#bits)
Wallclock time	$time(e)$	floating point	32/64
Sequence number	$seq(e)$	integer	32
Type of MPI call	$call(e)$	integer	8
Message data size	$size(e)$	integer	32
Message data address	$address(e)$	integer	64
Message tag	$tag(e)$	integer	32
Own rank	$rank(e)$	integer	32
Partner rank	$partner(e)$	integer	32
Callsite ID	$csite(e)$	integer	16
Program region	$region(e)$	integer	8

Fig. 3.1: Components of an event signature function.

Our goal for performance observation is to get an event inventory of an application (i.e., understand the events that happened and their characteristics) by associating performance data (number of occurrences, statistics on the duration) with event signatures. If the signature includes *time*(), we essentially have a model for event tracing because the chronology of events can be reconstructed from the stored signatures. If $time()$ and $seq()$ are not included in the signature, we have a model for profiling.

IPM is a profiling tool and for efficiency reasons we would like to keep the signature space much smaller than the event space ($|E| >> |S|$). In this case, the signature function will not be injective in general (many events can have the same signature) and performance data can be envisioned as a table indexed by the signature, with a number of columns for the statistics we are interested in. In IPM we implement this indexing using a hashtable resident in memory. The hash keys are 64 or 128 bits long and the timing statistics consume approximately 20 bytes per entry.

Upon program termination IPM writes a banner report and a log file in XML format. The banner contains the most important data about the program run in ASCII text and the log file provides the full information contained in the hashtable. In a post-processing step the XML is parsed and a profiling report is generated in HTML format. The contents of IPM's banner information and the full profiling report are discussed in the next section.

3.3 Performance Analysis with IPM

The most basic output IPM provides is a banner containing essential metrics which is written immediately after program termination. An example banner is shown in Fig. 3.2.

```
##IPM######################################################################
#
# command   : ./su3_rmd
# host      : nid03510                    mpi_tasks : 1024 on 86 nodes
# start     : Fri Nov 27 14:40:15 2009    wallclock : 20.57 sec
# stop      : Fri Nov 27 14:40:35 2009    %comm     : 55.69
# mem [GB]  : 108.98                      gflop/sec : 496.92
#
#           :      [total]        <avg>         min          max
# wallclock :     20961.67        20.47       20.42        20.57
# MPI calls :   1567890443      1531143     1531143      1531154
# MPI time  :     11673.29        11.40       11.26        11.94
# MPI [%]   :                     55.69       55.05        58.05
# mem [GB]  :       108.98         0.11        0.09         0.11
#
###########################################################################
```

Fig. 3.2: The default banner provided by IPM upon program termination contains a number of important high level metrics.

The banner provides general information about the executed job, such as the start and stop date and time, the number of MPI processes used and the number of different SMP nodes these processes where executing on. The next entries are the wallclock duration of the job in seconds and the percentage of overall time spent in MPI calls (`%comm`). If IPM has been installed on a system where these metrics can be acquired it also provides the total memory used by the application (`mem [GB]`) and the achieved floating point rate (`gflop/sec`).

The lower part of the banner provides information about the distribution of key metrics across the MPI ranks. For each of wallclock execution time, the time in MPI calls, the number of MPI calls, the percentage of time in MPI and the DRAM memory used, this section provides the sum, average, minimum and maximum values.

A further section of the banner (not shown) is included if a full banner is requested by setting the `IPM_BANNER` environment variable to `full`. This section provides a listing of the individual most time-consuming events stored in the hashtable, and their distribution over the MPI ranks.

In addition to the text-based banner, IPM writes the full profiling data to an XML-based log file. Parallel MPI file I/O is used at high concurrencies to speed up the creation of this log file. A parser script is provided with IPM that reads the XML log file and produces an HTML page of the full profiling report which also includes charts and graphs to visualize the data.

Among others, the HTML profiling page contains these entries:

- The information contained in the text-based banner is reproduced in a table on top of the profiling report.
- A pie chart (Fig. 3.3a) displays the breakdown of the total MPI time into the various contributing MPI calls such as `MPI_Allreduce` or `MPI_Wait`.
- For each monitored hardware counter event, the minimum, maximum and average values are displayed as well as the location (rank) of where the minimum and maximum values are achieved.
- A load balance line graph showing the consumed DRAM memory, floating point rate, and wallclock time. The horizontal axis is the rank dimension and the graphs are available both in sorted (by memory, flops, wallclock time), as well as unsorted (natural rank order) variant.
- A stacked load balance graph shows the breakdown of the MPI time into individual MPI calls over the rank dimension. An example for this graph (sorted by time) is shown in Fig. 3.3b. This type of display is especially useful to identify and locate load imbalance situations.
- Cumulative distribution graphs as the one shown in Fig. 3.3c provide an understanding of the message size distribution of an application. The horizontal axis is the buffer size n used in the operation and the vertical axis denotes how many calls have had a buffer size smaller or equal to n.
- A similar cumulative distribution graph is also provided where the accumulation is not performed over the number of calls but the time spent in the messaging operation instead.
- A communication topology graph as shown in Fig. 3.3d. This graph shows the amount of data exchanged between a pair of processes. The sending process is depicted on the horizontal axis, the receiving process is shown on the vertical axis.

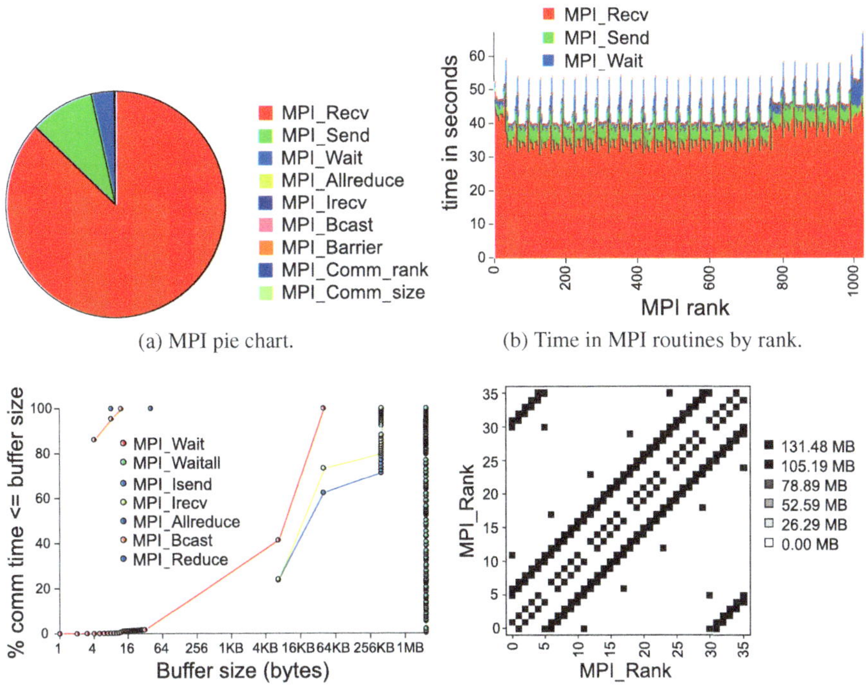

(a) MPI pie chart.

(b) Time in MPI routines by rank.

(c) Cumulative distribution graph of buffer sizes.

(d) Communication topology graph.

Fig. 3.3: Some of the performance data displays provided by IPM.

3.4 Example Scaling Study with IPM

This section describes an example scalability study using IPM. The application studied is MILC [6], a code to compute properties of elementary particles subject to the strong force as described by quantum chromodynamics (QCD). The computational domain is a 4D space-time lattice. Most of the computational work is spent doing a conjugate gradient inversion of the fermion Dirac operator. This entails repeated calls to a kernel that multiplies a large sparse matrix by a vector. Communication is mostly from point-to-point exchange of ghost-cell data on the 4D lattice. MILC was run with the medium problem size from the NERSC SSP (Sustained System Performance) benchmark set [8]. It performs 1375 conjugate gradient iterations on a 32^4 lattice.

The scaling study was performed on the Kraken Cray XT5 supercomputer at the National Institute for Computational Science (NICS) in Oakridge, Tennessee. At the time of writing this article Kraken was a dual-socket, 6-core AMD Opteron machine (12 cores total per node). Each processor is clocked at 2.6 GHz has 512

KB private second level cache, 6 MB shared L3 cache and each node has 12 GB of main memory.

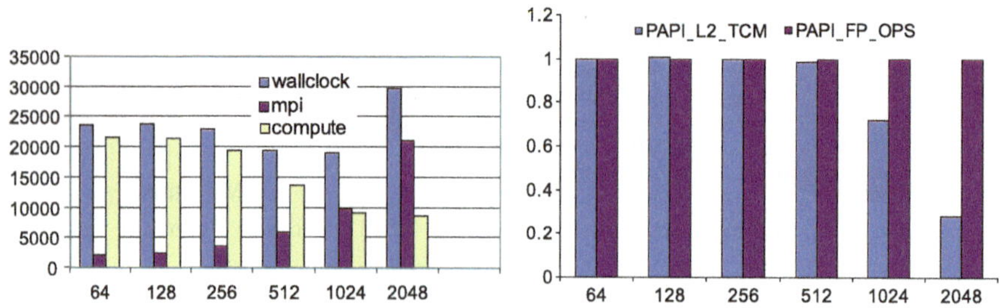

(a) Scaling of MILC. (b) Relative hardware counter scaling.

Fig. 3.4: MILC scaling study on Kraken.

We ran the program with 64, 128, 256, 512, 1024, and 2048 MPI processes and the graph in Fig. 3.4a shows the summed wallclock time (leftmost bar) for each run. Since this is a strong scaling study (the workload remains constant as the concurrency is increased), ideal scaling would be represented by constant summed wallclock times. For MILC, the summed wallclock execution time stays almost constant at 23,000 seconds for 64, 128, 256 processes, then drops to about 19,000 seconds for 512 and 1024, and then increases to about 30,000 seconds at 2048 processes.

Fig. 3.4a also shows the breakdown of the overall wallclock time into time spent in MPI operations and "compute" time (i.e., wallclock - MPI time). This breakdown sheds light on the nature of the wallclock scaling. Evidently the time in MPI (middle bar) increases dramatically as we increase the concurrency, from 2,016 seconds at 64 processes to 21,000 seconds at 2048 processes, but this increase is masked by a decrease in computation time (rightmost bar), most notably from 256 to 512 to 1024 MPI processes.

The most likely explanation for this super-linear speedup is the increased overall cache size at high concurrencies. At some point the data set will fit into the caches and cause less cache misses and traffic on the memory subsystem, leading to shorter execution times than could be expected from just the increase of the computing capacity alone. To test this hypothesis for MILC, we configured IPM to collect hardware counter events using PAPI [9], [1]. We measured the total number of floating point instructions (`PAPI_FP_OPS`) and the total number of level 2 cache misses (`PAPI_L2_TCM`). Figure 3.4b shows a chart of the relative scaling of the measured counter values as we increase the number of MPI processes. All values are normalized to the measured value at 64 processes. The number of floating point operations stays constant, since the same amount of work is performed in each run but the number of L2 cache misses decreases dramatically at 1024 and 2048 processes as the data set starts to fit the L2 cache.

By analyzing some of the basic metrics provided by IPM we were able to understand the scaling of MILC to a first order by separating out the contributions of messaging and computing. The next step would be to analyze the contribution of individual MPI calls (collective and point-to-point) and to consider how the message size distribution is changing and how the application might shift from being bandwidth bound to being latency bound. Both goals can be achieved readily using information provided in IPM's HTML profiling report.

3.5 Related Work

There are a number of other performance analysis tools, both employing tracing and profiling measurement techniques. Vampir [7], [2] is a trace collector and a trace visualizer for MPI and OpenMP applications. TAU [5], [11] is an extensive toolset for profiling and tracing of MPI/OpenMP applications and some other programming models. Automated performance analysis is the focus of other recent tools such as Periscope [3], [10] KOJAK [12], and Scalasca [4].

Compared to most of the aforementioned tools, IPM is focused more on giving the user a number of key metrics in a straightforward and easy way with low overheads and it places less emphasis on a detailed drill down into the application structure.

3.6 Conclusion and Outlook

We have introduce IPM, an Integrated Performance Monitoring framework with very low overhead. IPM can be used to derive essential key metrics for application characteristics such as percent of time spent in communication operations, imbalances in program regions or MPI calls, and the communication topology in a straightforward way.

For the path ahead, several improvements for IPM are under active investigation and development. We plan to add threading support to IPM for monitoring OpenMP and Pthreads applications. For file I/O a vertically integrated monitoring stack will allow us to monitor file operations at several layers (from the user's view to the underlying networking fabric) and will enable us gain novel insight into the interaction of storage, networking, and computation.

References

1. Shirley Browne, Jack Dongarra, N. Garner, G. Ho, and Philip J. Mucci. A portable programming interface for performance evaluation on modern processors. *Int. J. High Perform. Comput. Appl.*, 14(3):189–204, 2000.

2. Holger Brunst and Bernd Mohr. Performance analysis of large-scale OpenMP and hybrid MPI/OpenMP applications with VampirNG. In *Proceedings of the First International Workshop on OpenMP (IWOMP 2005)*, Eugene, Oregon, USA, May 2005.
3. Karl Fürlinger and Michael Gerndt. Periscope: Performance analysis on large-scale systems. *InSiDE – Innovatives Supercomputing in Deutschland (Featured Article)*, 3(2, Autumn):26–29, 2005.
4. Markus Geimer, Felix Wolf, Brian J. N. Wylie, and Bernd Mohr. Scalable parallel trace-based performance analysis. In *Proceedings of the 13th European PVM/MPI Users' Group Meeting on Recent Advances in Parallel Virtual Machine and Message Passing Interface (EuroPVM/MPI 2006)*, pages 303–312, Bonn, Germany, 2006.
5. Allen D. Malony and Sameer S. Shende. Performance technology for complex parallel and distributed systems. pages 37–46, 2000.
6. MILC website, `http://physics.indiana.edu/~sg/milc.html`.
7. Wolfgang E. Nagel, Alfred Arnold, Michael Weber, Hans-Christian Hoppe, and Karl Solchenbach. VAMPIR: Visualization and analysis of MPI resources. *Supercomputer*, 12(1):69–90, 1996.
8. Sustained system performance benchmarks at NERSC, `http://www.nersc.gov/projects/ssp.php`.
9. PAPI web page: `http://icl.cs.utk.edu/papi/`.
10. Periscope project homepage `http://wwwbode.cs.tum.edu/~gerndt/home/Research/PERISCOPE/Periscope.htm`.
11. Sameer S. Shende and Allen D. Malony. The TAU parallel performance system. *International Journal of High Performance Computing Applications, ACTS Collection Special Issue*, 2005.
12. Felix Wolf and Bernd Mohr. Automatic performance analysis of hybrid MPI/OpenMP applications. In *Proceedings of the 11th Euromicro Conference on Parallel, Distributed and Network-Based Processing (PDP 2003)*, pages 13–22. IEEE Computer Society, February 2003.

Chapter 4
Recent Developments in the Scalasca Toolset

Markus Geimer, Felix Wolf, Brian J. N. Wylie, Daniel Becker, David Böhme, Wolfgang Frings, Marc-André Hermanns, Bernd Mohr, and Zoltán Szebenyi

Abstract The number of processor cores on modern supercomputers is increasing from generation to generation, and as a consequence HPC applications are required to harness much higher degrees of parallelism to satisfy their growing demand for computing power. However, writing code that runs efficiently on large processor configurations remains a significant challenge. The situation is exacerbated by the rising number of cores imposing scalability demands not only on applications but also on the software tools needed for their development.
To address this challenge, Jülich Supercomputing Centre creates software technologies aimed at improving the performance of applications running on leadership-class systems. At the center of our activities lies the development of Scalasca, a performance-analysis tool that has been specifically designed for large-scale systems and that allows the automatic identification of harmful wait states in applications running on hundreds of thousands of processors. In this article, we review recent developments in the open-source Scalasca toolset, highlight research activities of the Scalasca team during the past two years and give an outlook on future work.

Markus Geimer, Felix Wolf, Brian J. N. Wylie, David Böhme, Wolfgang Frings, Bernd Mohr, Zoltán Szebenyi
Jülich Supercomputing Centre,
Forschungszentrum Jülich, 52425 Jülich, Germany
e-mail: {m.geimer, b.wylie, d.boehme, w.frings, b.mohr, z.szebenyi}@fz-juelich.de

Felix Wolf, Daniel Becker, Marc-André Hermanns
German Research School for Simulation Sciences, 52062 Aachen, Germany
e-mail: {f.wolf, d.becker, m.a.hermanns}@grs-sim.de

Felix Wolf, David Böhme, Zoltán Szebenyi
RWTH Aachen University, 52056 Aachen, Germany

M.S. Müller et al. (eds.), *Tools for High Performance Computing 2009*,
DOI 10.1007/978-3-642-11261-4_4, © Springer-Verlag Berlin Heidelberg 2010

4.1 Introduction

Supercomputing is a key technology pillar of modern science and engineering, indispensable to solve critical problems of high complexity. The extension of the ESFRI road map to include a European supercomputer infrastructure in combination with the creation of the PRACE consortium acknowledges that the requirements of many critical applications can only be met by the most advanced custom-built large-scale computer systems. However, as a prerequisite for their productive use, the HPC community needs powerful and robust software development tools. These would not only help improve the scalability characteristics of scientific codes and thus expand their potential, but also allow domain scientists to concentrate on the underlying models rather than to spend a major fraction of their time tuning their application for a particular machine.

As the current trend in microprocessor development continues, this need will become even stronger in the future. Facing increasing power dissipation and little instruction-level parallelism left to exploit, computer architects are realizing further performance gains by using larger numbers of moderately fast processor cores rather than by increasing the speed of uni-processors. As a consequence, supercomputer applications are being required to harness much higher degrees of parallelism in order to satisfy their growing demand for computing power. With an exponentially rising number of cores, the often substantial gap between peak performance and the performance level actually sustained by production codes is expected to widen even further. Finally, increased concurrency levels place higher scalability demands not only on applications but also on parallel programming tools. When applied to larger numbers of cores, familiar tools often cease to work in a satisfactory manner (e.g., due to escalating memory requirements, failing renditions, or limited I/O performance).

To overcome this challenge, Jülich Supercomputing Centre creates software technologies aimed at improving the performance of applications running on leadership-class systems with hundreds of thousands of cores. At the center of our activities lies the development of Scalasca [1, 2], an open-source performance-analysis tool that has been specifically designed for large-scale systems, which allows the automatic identification of harmful wait states in applications running on very large processor configurations.

In this article, we give an overview of Scalasca and highlight research accomplishments of the Scalasca team during the past two years, focusing on the analysis of hybrid applications, the detection of wait states, and the characterization of time-dependent behavior. The latter two examples address the scalability of Scalasca regarding both the number of processes and the length of execution, respectively.

4.2 Scalasca Overview

Scalasca supports measurement and analysis of MPI applications written in C, C++ and Fortran on a wide range of current HPC platforms [3]. Hybrid codes making use of basic OpenMP features in addition to message passing are also supported. Figure 4.1 shows the basic analysis workflow supported by Scalasca. Before any performance data can be collected, the target application must be instrumented and linked to the measurement library. When running the instrumented code on the parallel machine, the user can choose between generating a summary analysis report ('profile') with aggregate performance metrics for individual function call paths and/or generating event traces recording individual runtime events from which a profile or time-line visualization can later be produced. Summarization is particularly useful to obtain an overview of the performance behavior and for local metrics such as those derived from hardware counters. Since traces tend to rapidly become very large [4], optimizing the instrumentation and measurement based on the summary report is usually recommended. When tracing is enabled, each process generates a trace file containing records for its process-local events. After program termination, Scalasca loads the trace files into main memory and analyzes them in parallel using as many processes as have been used for the target application itself. During the analysis, Scalasca searches for wait states and related performance properties, classifies detected instances by category, and quantifies their significance. The result is a wait-state report similar in structure to the summary report but enriched with higher-level communication and synchronization inefficiency metrics. Both summary and wait-state reports contain performance metrics for every measured function call path and process/thread which can be interactively examined in the provided analysis report explorer.

4.3 Analysis of Hybrid MPI/OpenMP Codes

Although message passing is still the predominant programming paradigm used in HPC, increasingly applications leverage OpenMP to exploit more fine-grained process-local parallelism, while communicating between processes using MPI. Support for such hybrid applications in the Scalasca 1.0 release consisted of serial trace analysis of merged traces using the EXPERT analyzer from the KOJAK toolkit [5]. Extended runtime summarization and automatic parallel trace analysis support incorporated in Scalasca 1.2 provide similar analyses of hybrid OpenMP/MPI applications, within the same Scalasca instrumentation, measurement collection and analysis, and presentation usage model [6].

The OPARI source-code preprocessor inserts instrumentation for OpenMP constructs and API calls, which deliver events to the OpenMP-aware measurement library. Call-path metrics are accumulated per OpenMP thread during measurement, and collated into a complete summary report during finalization. Trace data is also analyzed in parallel with an analyzer thread for each OpenMP thread, and subse-

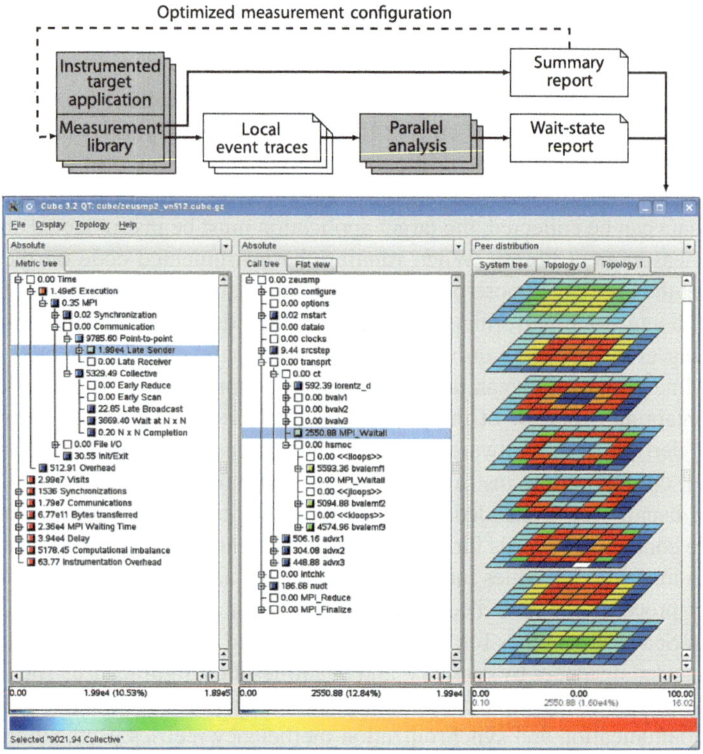

Fig. 4.1: Schematic overview of the performance data flow in Scalasca. Grey rectangles denote programs and white rectangles with the upper right corner turned down denote files. Stacked symbols denote multiple instances of programs or files running or being processed in parallel. The GUI shows the distribution of performance metrics (left pane) across the call tree (middle pane) and the process topology (right pane).

quently collated into a similar pattern report. Event timestamp correction can also be applied to the trace data of OpenMP thread teams when logical consistency violations are encountered in MPI events due to unsynchronized clocks. Specific OpenMP metrics are calculated and presented alongside serial and MPI metrics in integrated analysis reports.

While the trace analysis currently remains restricted to fixed-size teams of OpenMP threads, runtime summarization identifies threads that have not been used during parallel regions. The associated time within the parallel region is distinguished as a *Limited parallelism* metric from the *Idle threads* time which includes time outside OpenMP parallel regions when only the master thread executes. This

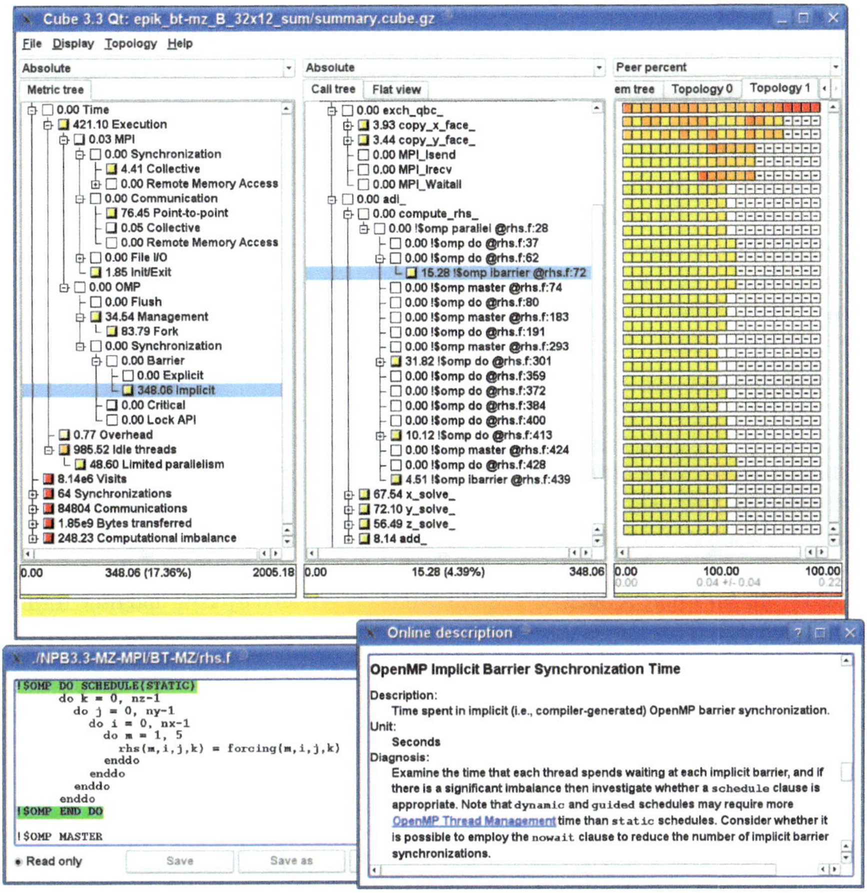

Fig. 4.2: Scalasca analysis report explorer display of a hybrid OpenMP/MPI NPB3.3 BT-MZ benchmark Class B execution on 32 Cray XT5 twin six-core compute nodes, showing OpenMP *Implicit Barrier Synchronization* time for a parallel loop in the `compute_rhs` routine (from lines 62 to 72 of source file rhs.f) broken down by thread. Higher metric values are shown darker and void thread locations in the topology pane are displayed in gray or with dashes.

matches the typical usage of dedicated HPC resources which are allocated for the duration of the parallel job, or threads which busy-wait occupying compute resources in shared environments. The number of OpenMP threads included in the measurement can be explicitly specified, defaulting to the number of threads for an unqualified parallel region when measurement commences: a warning is provided if subsequent `omp_set_num_threads` calls or `num_threads` clauses result in additional threads not being included in the measurement experiment.

Figure 4.2 shows a Scalasca analysis report from a hybrid OpenMP/MPI NAS NPB3.3 Block Triangular Multi-Zone (BT–MZ) benchmark [7] Class B execution in a Cray XT5 partition consisting of 32 compute nodes, each with two six-core Opteron processors. One MPI process was started on each of the compute nodes, and OpenMP threads run within each SMP node. In an unsuccessful attempt at load balancing by the application, more than 12 OpenMP threads were created by the first 6 MPI ranks (shown at the top of the topological presentation in the right pane), and 20 of the remaining ranks used fewer than 12 OpenMP threads. While the 49 seconds of *Limited parallelism* time for the unused cores represent only 2% of the allocated compute resources, half of the total time is wasted by *Idle threads* while each process executes serially, including MPI operations done outside of parallel regions by the master thread of each process. Although the exclusive *Execution* time in local computation is relatively well balanced on each OpenMP thread, the over-subscription of the first 6 compute nodes manifests as excessive *Implicit Barrier Synchronization* time at the end of parallel regions (as well as additional OpenMP *Thread Management* overhead), and higher *MPI Point-to-point Communication* time on the other processes is then a consequence of this. When over-subscription of cores is avoided, benchmark execution time is reduced by one third (with *MPI* time reduced by 52%, *OMP* time reduced by 20% and time for *Idle threads* reduced by 55%).

4.4 Scalable Wait-State Analysis

In message-passing applications, processes often require access to data provided by remote processes, making the progress of a receiving process dependent upon the progress of a sending process. Collective synchronization is similar in that its completion requires each participating process to have reached a certain point. As a consequence, a significant fraction of the communication and synchronization time can often be attributed to wait states, for example, as a result of an unevenly distributed workload. Especially when trying to scale applications to large process counts, such wait states can present severe challenges to achieving good performance.

4.4.1 Scalability

After the target application has terminated and the trace data have been flushed to disk, the trace analyzer is launched with one analysis process per (target) application process and loads the entire trace data into its distributed memory address space. Future versions of Scalasca may exploit persistent memory segments to pass the trace data to the analysis stage without involving any file I/O. While traversing the traces in parallel, the analyzer performs a replay of the application's original communication behavior [8]. During the replay, the analyzer identifies wait states

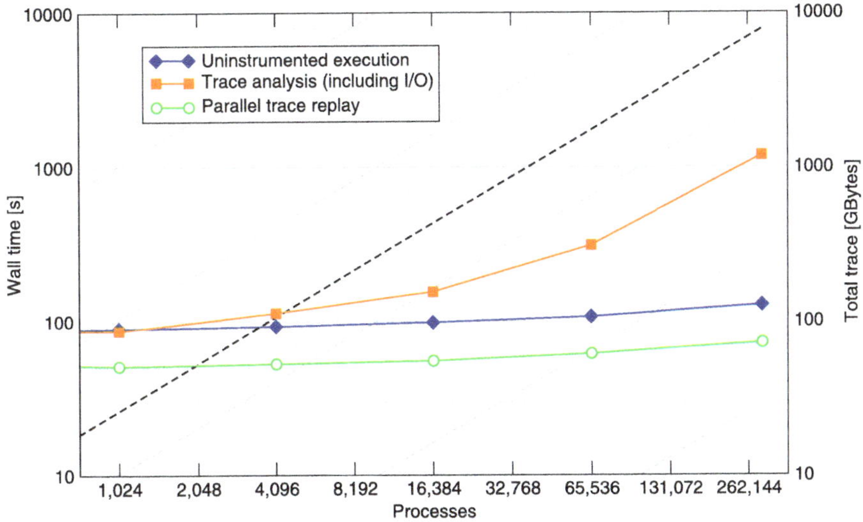

Fig. 4.3: Scalability of wait-state search for the ASCI benchmark application SWEEP3D on the JUGENE Blue Gene/P. The graph charts wall-clock execution times for the uninstrumented application and for the analyses of trace files generated by the instrumented version with varying numbers of processes. The time needed for the trace analysis replay is shown as well as that for the entire parallel analysis (including loading the traces and collating the analysis report). The black dashed line shows the linear increase in total trace size in GBytes.

in communication operations by measuring temporal differences between local and remote events after their timestamps have been exchanged using an operation of similar type. Since trace processing capabilities (i.e., processors and memory) grow proportionally with the number of application processes, we can achieve good scalability at previously intractable scales. Recent scalability improvements allowed us to perform trace analyses of execution measurements with up to 294,912 processes (Figure 4.3).

4.4.2 Improvement of Trace-Data I/O

Parallel applications often store data in multiple task-local files, for example, to remember checkpoints, to circumvent memory limitations, or to record trace data in the case of the Scalasca toolset. When operating at very large processor configurations, such applications often experience scalability limitations when the simultaneous creation of thousands of files causes metadata-server contention or simply when large file counts complicate file management or operations on those files destabilize the file system. In this context, a generic parallel I/O library called SIONlib has been developed which addresses this problem by transparently mapping a large number of task-local files onto a small number of physical files via internal metadata handling and block alignment to ensure high performance. While requiring only minimal source code changes, SIONlib significantly reduces file creation overhead and simplifies the resulting file handling, offering even the chance to achieve superior read and write performance via optimized I/O operations [9]. For the Scalasca trace collection and analysis of 294,912 processes shown in Figure 4.3, SIONlib was able to reduce the time to create the experiment archive directory and trace files from 86 minutes (for individual files) down to 10 minutes (for one file for each of the 576 BG/P I/O nodes),

4.4.3 Analysis of MPI-2 Remote Memory Access Operations

In our earlier work, we already defined wait-state patterns for MPI-2 *Remote Memory Access* (RMA) communication and synchronization, although still based on a serial trace-analysis scheme with limited scalability [10]. Taking advantage of Scalasca's scalable trace-analysis approach, we recently extended our parallel trace analyzer to detect these wait states. Using the programming paradigm of the target application, RMA-related communication and synchronization inefficiencies are now detected by exchanging data via RMA operations. In this way, we successfully performed analyses of RMA-based applications running with up to 8,192 processes. [11]

4.4.4 Delay Analysis

In general, the temporal or spatial distance between cause and symptom of a performance problem constitutes a major difficulty in deriving helpful conclusions from performance data. So just knowing the locations of wait states in the program is often insufficient to understand the reason for their occurrence. We are currently extending our replay-based wait-state analysis in such a way that it attributes the waiting times to their root causes. The root cause, which we call a *delay*, is an interval during which a process performs some additional activity not performed by its peers, for example as a result of insufficiently balancing the load. [12]

4.4.5 Evaluation of Optimization Hypotheses

Excess workload identified as root cause of wait states usually cannot simply be removed. To achieve a better balance, optimization hypotheses drawn from a delay analysis typically propose the redistribution of the excess load to other processes instead. However, redistributing workloads in complex message-passing applications can have intricate side-effects that may compromise the expected reduction of waiting times. Given that balancing the load statically or even introducing a dynamic load-balancing scheme constitute major code changes, they should ideally be performed only if the prospective performance gain is likely to materialize. Our goal is therefore to automatically predict the effects of redistributing a given delay without altering the application itself and to determine the savings that can be realistically hoped for. Since the effects of such changes are hard to quantify analytically, we simulate these changes via a real-time replay of event traces after they have been modified to reflect the redistributed load. [13, 14]

4.4.6 Configurable Source-Code Instrumentation

Proper instrumentation is an essential prerequisite for producing reliable performance analysis results with the Scalasca toolset. We therefore extended our instrumentation capabilities to leverage the generic and configurable source-code instrumentation component we developed in collaboration with the University of Oregon based on PDT and the TAU instrumentor. [15] This component provides flexible instrumentation specification capabilities, reducing the need to filter performance events at runtime and, thus, further reducing the measurement overhead.

4.5 Analysis of Time-Dependent Behavior

As scientific parallel applications simulate the temporal evolution of a system their progress occurs via discrete points in time. Accordingly, the core of such an application is typically a loop that advances the simulated time step by step. However, the performance behavior may vary between individual iterations, for example, due to periodically re-occurring extra activities [16] or when the state of the computation adjusts to new conditions in so-called adaptive codes [17].

4.5.1 Observing Individual Iterations

To study the time-dependent behavior, Scalasca is being equipped with iteration instrumentation capabilities (corresponding to TAU dynamic timers [18]) that allow the distinction of individual iterations both in runtime summaries and in event traces. Moreover, to simplify the understanding of the resulting time-series data, we are implementing several display tools including iteration graphs with minimum, median,

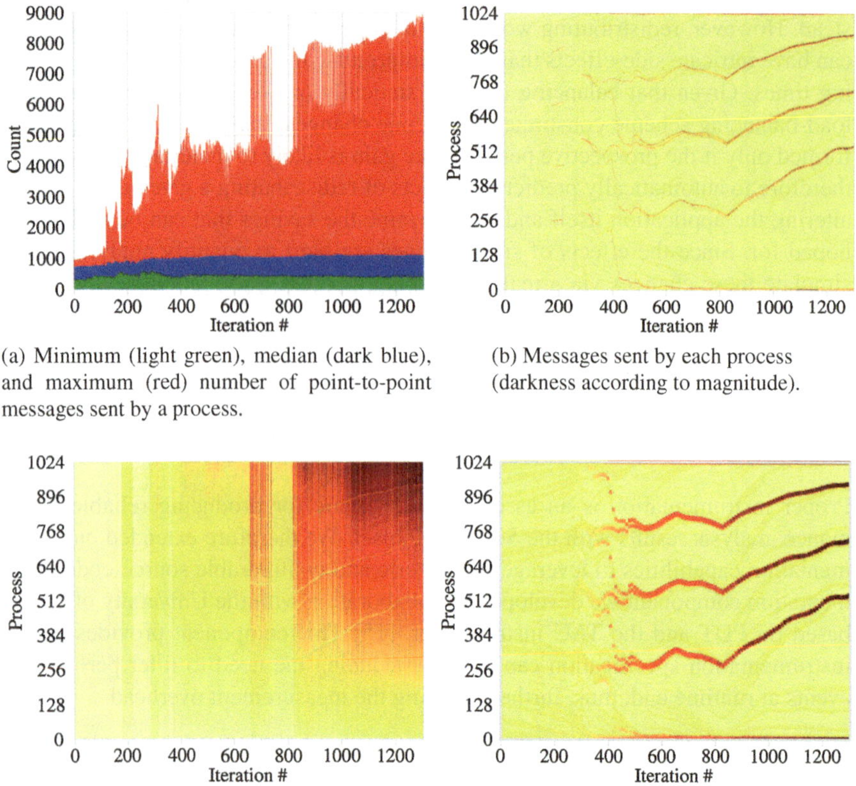

(a) Minimum (light green), median (dark blue), and maximum (red) number of point-to-point messages sent by a process.

(b) Messages sent by each process (darkness according to magnitude).

(c) Late Sender waiting time of a process.

(d) Particles held by each process.

Fig. 4.4: Gradual development of a performance problem over 1,300 timesteps of the PEPC application execution on 1,024 processors of Blue Gene.

and maximum representation (Figure 4.4a) as well as value maps to cover the full <process/thread, iteration> space for a given performance metric (Figure 4.4b).

Using prototype implementations of these new tools, we evaluated the performance behavior of the SPEC MPI2007 benchmark suite on the IBM SP p690 cluster JUMP, observing a large variety of complex temporal characteristics ranging from gradual changes and sudden transitions of the base-line behavior to both periodically and irregularly occurring peaks, including noise that varies from measurement to measurement [19]. Moreover, problems with several benchmarks that limited their scalability (sometimes to only 128 processes) were identified, such as distributing initialization data via broadcasts in 113.GemsFDTD and insufficiently large data sets for several others. Even those codes that apparently scaled well contained considerable quantities of waiting time, indicating possible opportunities for performance and scalability improvement through more effective work distributions or bindings of processes to processors.

Another real-world code with a substantially time-varying execution profile is the PEPC [20] particle simulation code, developed at Jülich Supercomputing Centre and the subject of an application liaison between the Scalasca and PEPC developer teams. The MPI code employs a parallel tree algorithm to efficiently calculate the forces the particles exert on each other and also includes a load-balancing mechanism that redistributes the computational load by shifting particles between processes. However, our analysis [21] revealed a severe and gradually increasing communication imbalance (Figure 4.4a). We found evidence that the imbalance was caused by a small group of processes with time-dependent constituency that sent large numbers of messages to all remaining processes (Figure 4.4b) in rank order, introducing *Late Sender* waiting times at processes with higher ranks (Figure 4.4c). Interestingly, the communication imbalance correlated very well with the number of particles "owned" by a process (Figure 4.4d), suggesting that the load-balancing scheme smoothes the computational load at the expense of communication disparities. Since the number of particles also influence the memory requirements of a process, we further conclude that the current behavior of concentrating particles at a small subset of processes may adversely affect scalability under different configurations. Work with the application developers to revise the load-balancing scheme and improve the communication efficiency is in progress.

4.5.2 Space-Efficient Time-Series Call-Path Profiling

While call-path profiling is an established method of linking a performance problem to the context in which it occurs, generating call-path profiles separately for thousands of iterations may exceed the available buffer space — especially when the call tree is large and more than one metric is collected. We therefore developed a runtime approach for the semantic compression of call-path profiles [22] based on incremental clustering of a series of single-iteration profiles that scales in terms of the number of iterations without sacrificing important performance details. Our approach has low runtime overhead by using only a condensed version of the profile data when calculating distances and accounts for process-dependent variations by making all clustering decisions locally.

4.6 Outlook

Besides further scalability improvements in view of upcoming systems in the range of several petaflops, we plan to extend Scalasca towards emerging programming models such as partitioned global address space languages and general-purpose GPU programming, which we expect to play a bigger role in the future. Moreover, to offer enhanced functionality and combine development efforts, we will integrate

Scalasca closer with related tools including Periscope [23], TAU [24], and Vampir [25].

Acknowledgement Financial support from the Helmholtz Association of German Research Centers through Grants VH-NG-118 and VH-VI-228 and the German Research Foundation through Grant GSC 111 is gratefully acknowledged. This research was supported by allocations of advanced computing resources provided by the John von Neumann Institute for Computing and US National Science Foundation for computations on the Jugene IBM Blue Gene/P at Jülich Supercomputing Centre and Kraken Cray XT5 at the US National Institute for Computational Sciences.

References

1. Jülich Supercomputing Centre: Scalasca. `http://www.scalasca.org/`.
2. Geimer, M., Wolf, F., Wylie, B.J.N., Ábrahám, E., Becker, D., Mohr, B.: The Scalasca performance toolset architecture. Concurrency and Computation: Practice and Experience, Proc. Workshop on Scalable Tools for High-End Computing (to appear) DOI: 10.1002/cpe.1556.
3. Wylie, B.J.N., Geimer, M., Wolf, F.: Performance measurement and analysis of large-scale parallel applications on leadership computing systems. Scientific Programming **16**(2-3) (2008) 167–181
4. Wolf, F., Freitag, F., Mohr, B., Moore, S., Wylie, B.J.N.: Large event traces in parallel performance analysis. In: Proc. 8th Workshop on Parallel Systems and Algorithms (PASA, Frankfurt/Main, Germany). Lecture Notes in Informatics, Gesellschaft für Informatik (March 2006) 264–273
5. Wolf, F., Mohr, B.: Automatic performance analysis of hybrid MPI/OpenMP applications. In: Proc. 11th Euromicro Conf. on Parallel Distributed and Network based Processing (Genoa, Italy), IEEE Computer Society (February 2003) 13–22
6. Wolf, F., Wylie, B.J.N., Ábrahám, E., Becker, D., Frings, W., Fürlinger, K., Geimer, M., Hermanns, M.A., Mohr, B., Moore, S., Pfeifer, M., Szebenyi, Z.: Usage of the SCALASCA toolset for scalable performance analysis of large-scale parallel applications. In: Proc. 2nd HLRS Parallel Tools Workshop (Stuttgart, Germany), Springer (July 2008) 157–167 ISBN 978-3-540-68561-6.
7. Van der Wijngaart, R.F., Jin, H.: NAS Parallel Benchmarks, Multi-Zone versions. Technical Report NAS-03-010, NASA Ames Research Center, Moffett Field, CA, USA (July 2003)
8. Geimer, M., Wolf, F., Wylie, B.J.N., Mohr, B.: A scalable tool architecture for diagnosing wait states in massively-parallel applications. Parallel Computing **35**(7) (2009) 375–388
9. Frings, W., Wolf, F., Petkov, V.: Scalable massively parallel I/O to task-local files. In: Proc. 21st ACM/IEEE SC Conf. (SC09, Portland, OR, USA). (November 2009)
10. Kühnal, A., Hermanns, M.A., Mohr, B., Wolf, F.: Specification of inefficiency patterns for MPI-2 one-sided communication. In: Proc. 12th Euro-Par (Dresden, Germany). Volume 4128 of Lecture Notes in Computer Science, Springer (2006) 47–62
11. Hermanns, M.A., Geimer, M., Mohr, B., Wolf, F.: Scalable detection of MPI-2 remote memory access inefficiency patterns. In: Proc. 16th European PVM and MPI Conference (EuroPVM/MPI, Espoo, Finland). Volume 5759 of Lecture Notes in Computer Science, Springer (September 2009) 31–41
12. Böhme, D., Geimer, M., Hermanns, M.A., Wolf, F.: Identifying the root causes of wait states in large-scale parallel applications. Technical Report AICES-2010-1, Aachen Institute for Advanced Study in Computational Engineering Science, RWTH Aachen University, Germany (January 2010)
13. Hermanns, M.A., Geimer, M., Wolf, F., Wylie, B.J.N.: Verifying causality between distant performance phenomena in large-scale MPI applications. In: Proc. 17th Euromicro Int'l Conf. on

Parallel, Distributed, and Network-Based Processing (PDP, Weimar, Germany), IEEE Computer Society (February 2009) 78–84
14. Böhme, D., Hermanns, M.A., Geimer, M., Wolf, F.: Performance simulation of non-blocking communication in message-passing applications. In: Proc. 2nd Workshop on Productivity and Performance (PROPER 2009, Delft, The Netherlands). (August 2009) (to appear).
15. Geimer, M., Shende, S.S., Malony, A.D., Wolf, F.: A generic and configurable source-code instrumentation component. In: Proc. 9th Int'l Conf. on Computational Science (ICCS, Baton Rouge, LA, USA). Volume 5545 of Lecture Notes in Computer Science, Springer (May 2009) 696–705
16. Kerbyson, D.J., Barker, K.J., Davis, K.: Analysis of the weather research and forecasting (WRF) model on large-scale systems. In: Proc. 12th Conference on Parallel Computing (ParCo, Aachen/Jülich, Germany). Volume 15 of Advances in Parallel Computing, IOS Press (September 2007) 89–98
17. Shende, S., Malony, A., Morris, A., Parker, S., de St. Germain, J.: Performance evaluation of adaptive scientific applications using TAU. In: Parallel Computational Fluid Dynamics — Theory and Applications. Elsevier (2006) 421–428
18. Malony, A.D., Shende, S.S., Morris, A.: Phase-based parallel performance profiling. In: Proc. 11th Conference on Parallel Computing (ParCo, Málaga, Spain). Volume 33 of NIC Series, John von Neumann Institute for Computing (September 2005) 203–210
19. Szebenyi, Z., Wylie, B.J.N., Wolf, F.: SCALASCA parallel performance analyses of SPEC MPI2007 applications. In: Proc. 1st SPEC Int'l Performance Evaluation Workshop (SIPEW, Darmstadt, Germany). Volume 5119 of Lecture Notes in Computer Science, Springer (June 2008) 99–123
20. Gibbon, P., Frings, W., Dominiczak, S., Mohr, B.: Performance analysis and visualization of the N-body tree code PEPC on massively parallel computers. In: Proc. 11th Conf. on Parallel Computing (ParCo, Málaga, Spain). Volume 33 of NIC Series, John von Neumann Institute for Computing (October 2005) 367–374
21. Szebenyi, Z., Wylie, B.J.N., Wolf, F.: Scalasca parallel performance analyses of PEPC. In: Proc. 1st EuroPar Workshop on Productivity and Performance (PROPER 2008, Las Palmas de Gran Canaria, Spain). Volume 5415 of Lecture Notes in Computer Science, Springer (August 2008) 305–314
22. Szebenyi, Z., Wolf, F., Wylie, B.J.N.: Space-efficient time-series call-path profiling of parallel applications. In: Proc. 21st ACM/IEEE SC Conference (SC09, Portland, OR, USA). (November 2009)
23. Technical University of Munich: Periscope. http://www.lrr.in.tum.de/~gerndt/home/Research/PERISCOPE/Periscope.htm.
24. University of Oregon: TAU. http://www.cs.uoregon.edu/research/tau/.
25. Technische Universität Dresden: Vampir. http://www.vampir.eu/.

Chapter 5
MUST: A Scalable Approach to Runtime Error Detection in MPI Programs

Tobias Hilbrich, Martin Schulz, Bronis R. de Supinski, and Matthias S. Müller

Abstract The Message-Passing Interface (MPI) is large and complex. Therefore, programming MPI is error prone. Several MPI runtime correctness tools address classes of usage errors, such as deadlocks or non-portable constructs. To our knowledge none of these tools scales to more than about 100 processes. However, some of the current HPC systems use more than 100,000 cores and future systems are expected to use far more. Since errors often depend on the task count used, we need correctness tools that scale to the full system size. We present a novel framework for scalable MPI correctness tools to address this need. Our fine-grained, module-based approach supports rapid prototyping and allows correctness tools built upon it to adapt to different architectures and use cases. The design uses P^nMPI to instantiate a tool from a set of individual modules. We present an overview of our design, along with first performance results for a proof of concept implementation.

5.1 Introduction

The Message Passing Interface (MPI) [1, 2] is the de-facto standard for programming HPC (High Performance Computing) applications. Even the first version of this interface offers more than 100 different functions to provide various types of data transfers. Thus MPI usage is error prone and debugging tools can greatly in-

Tobias Hilbrich
GWT-TUD GmbH, Chemnitzer Str. 48b, 01187 Dresden, Germany
e-mail: tobias.hilbrich@zih.tu-dresden.de

Martin Schulz, Bronis R. de Supinski
Lawrence Livermore National Laboratory, Livermore, CA 94551, USA
e-mail: {schulzm, bronis}@llnl.gov

Matthias S. Müller
Center for Information Services and High Performance Computing (ZIH), Technische Universität Dresden, D-01062 Dresden, Germany
e-mail: matthias.mueller@tu-dresden.de

M.S. Müller et al. (eds.), *Tools for High Performance Computing 2009*,
DOI 10.1007/978-3-642-11261-4_5, © Springer-Verlag Berlin Heidelberg 2010

crease MPI programmers' productivity. Many types of errors can occur with MPI usage including invalid arguments, errors in type matching, race conditions, deadlocks and portability errors. Existing tools that detect some of these errors use one the following three approaches: static source code analysis, model checking or runtime error detection.

Runtime error detection is usually the most practical of these approaches for tool users, since it can be deployed transparently and avoids the potentially exponential analysis time of static analysis or model checking. However, these tools are generally limited to the detection of errors that occur in the executed control flow of the application and, thus, may not identify all potential errors. Several runtime error detection tools for MPI exist; however, our experience is that none of these tools covers all types of MPI errors. Further, none is known to scale to more than about 100 processes. With current systems that utilize more than 100,000 cores it is becoming increasingly difficult to apply these tools, even to small test cases.

This paper presents MUST, a new approach to runtime error detection in MPI applications. It draws upon our previous experience with the existing tools Marmot [3] and Umpire [4] and is specifically designed to overcome the scalability limitations of current runtime detection tools while facilitating the implementation of additonal detection routines. MUST relies on a fine grain design in the form of modules that are loaded into P^nMPI [5]. The next section will present the experiences and issues that we discovered during our development of Marmot and Umpire. Section 5.3 presents the goals and general design ideas of MUST, while Section 5.4 covers several key design details of MUST. In Section 5.5 we present initial experimental results with a proof of concept implementation of the MUST design. Finally, Sections 5.6 and 5.7 present related work and our conclusions.

5.2 Experiences from Marmot and Umpire

This section presents insights into our two predecessor MPI correctness checking tools: Marmot [3] and Umpire [4]. Marmot provides a wide range of local and global checks and offers good usability and integration into several other tools. Umpire's strength is a runtime deadlock detection algorithm that detects all actual deadlocks in MPI-1.2 as well as some potential deadlocks on alternate execution paths. While both tools have been very successful and have helped users debug their codes, they both are first generation MPI checker tools and have inherent limitations, upon which we focus in the following.

In particular, our analysis focuses on two things: first, the communication system for MPI trace records; second, the separation of tool internal infrastructure and the actual correctness checks. The communication system is necessary for checks (e.g., deadlock detection or type matching) that require global knowledge of MPI calls, i.e., data from more than one process. Thus, such checks require a system to communicate records for MPI calls. The separation of tool internal infrastructure and the actual correctness checks is important in order to enhance existing checks and to add further correctness checks that are used for new features or new versions of the MPI standard. We first analyze these aspects for Marmot and then cover Umpire.

5.2.1 Marmot

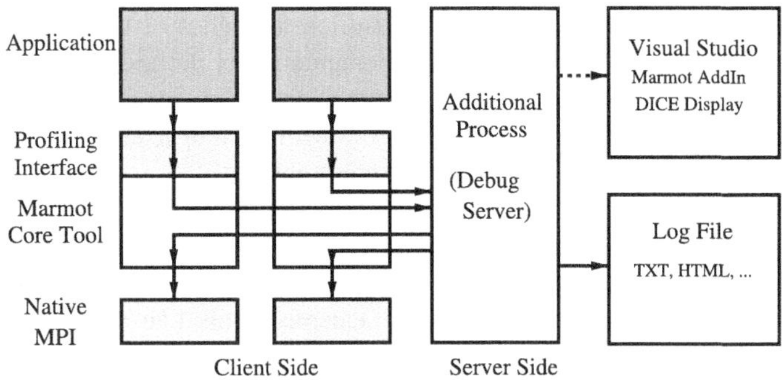

Fig. 5.1: Marmot trace communication design.

Marmot is an MPI runtime checker written in C++ that covers MPI-1.2 and parts of MPI-2. Its communication system is sketched in Figure 5.1. Marmot's MPI wrappers intercept any MPI call issued by the application. Marmot then performs two steps before executing the actual MPI call: first, it checks for correctness of the MPI call locally; second, it sends a trace record for this MPI call to the "DebugServer", a global manager process. The application process continues its execution only after it receives a ready-message from the *DebugServer*. As a result, it is guaranteed that all non-local checks executed at the *DebugServer*, as well as all local, are finished before the actual MPI call is issued. This synchronous checking ensures that all errors are reported before they can actually occur, which removes the need to handle potential application crashes. The *DebugServer* also executes a timeout based deadlock detection. While this approach can detect many deadlocks, it can lead to false positives. Also, it is not possible to highlight the MPI calls that lead to a deadlock with this strategy. Additionally, the *DebugServer* performs error logging in various output formats and can send error reports via TCP socket communication to arbitrary receivers. The main disadvantage of this synchronous or blocking communication system is its high impact on application performance. In particular, the runtime overhead increases significantly as the number of MPI processes increases since the *DebugServer* is a centralized bottleneck. Also, the blocking communication with the *DebugServer* can lead to high latency even at small scales, which – especially for latency bound applications – is a disadvantage.

The separation of tool internal infrastructure and the actual MPI correctness checks is not well solved for Marmot. It uses one C++ class for each MPI call and uses multiple abstract classes to build a hierarchy for all MPI calls. Checks are implemented as methods of these classes and are called before the PMPI call is issued. This has two disadvantages: First, checks for one MPI call are often distributed to multiple objects making it hard to determine which checks are used for a certain MPI call. Second, our experience with Marmot shows that there is no reasonable hierarchy for MPI calls that also builds a good hierarchy for all the different types of checks. Thus, many checks in Marmot are either implemented in very abstract classes or are implemented in multiple branches of the object hierarchy, which leads to code redundancy. The implementation of the checks uses a multitude of static variables that are stored in the more abstract classes of the hierarchy. These variables represent state information for the MPI system leading to checks being very tightly coupled with Marmot's class hierarchy.

The development of Marmot occurred concurrently with multiple workshops on parallel programming tools that included hands-on sessions. The experiences from these workshops guided the development of Marmot. One of the commonly asked-for features are integrations into widely accepted tools like debuggers, IDEs, or performance tools. As a result, Marmot provides multiple usability enhancing tools and integrations that help users in applying the tool. These efforts help new users to apply the tool easily, which is an important factor for the success of Marmot.

5.2.2 *Umpire*

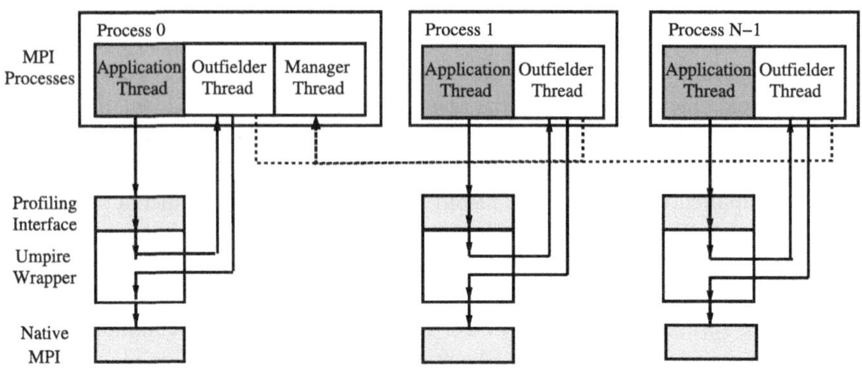

Fig. 5.2: Umpire trace communication design.

The MPI correctness checker Umpire is written in C and focuses on non-local MPI checks. It executes both a centralized deadlock detection and type matching at a central manager. Figure 5.2 sketches the trace transfer that is implemented in

Umpire. The first difference to Marmot is that Umpire spawns extra threads for each MPI process. It spawns an "outfielder" thread for all processes. In addition, it spawns a "manager" thread on one process (usually process 0). The *outfielder* thread asynchronously transfers trace records to the centralized manager, which is executed on the *manager* thread.

Similarly to Marmot, Umpire's wrappers intercept any MPI call issued by the application. However, Umpire minimizes immediate application perturbation. The application thrad only builds a trace record for the MPI call, which it transfers to the *outfielder* thread of that process through shared memory. Each *outfielder* thread aggregates the trace records that it receives and sends them to the *manger* thread. This send happens if the buffer limit is exceeded or when a timeout occurs. This communication is implemented with either MPI or shared memory depending on the system architecture. Umpire's communication system is designed to incur low runtime overhead, which is achieved with the asynchronous transfer of trace records to the central manager. Due to the asynchronous design, the central manager is no longer a bottleneck. However, it still limits performance since it must analyze trace records of all processes. Further, performance tests with Umpire show that the efficiency of the asynchronous transfer depends highly on the interleaving of the communication of the application and the MPI communication of the *outfielder* threads [6].

As with Marmot, the separation of internal infrastructure and correctness checks is incomplete with Umpire. The checks that are executed at the centralized manager are tightly coupled to a large structure that represents internal state as well as MPI state. All checks are directly coupled to this structure. Also, some of the different checks of the central manager are dependent on each other and need to use internal data from each other. This applies to a smaller extent to local checks which tend to need less state information. Umpire currently only implements a small number of local checks. Additional local checks may be added by extending the wrapper generation of Umpire, since checks can be issued in the wrappers or other generated files.

5.3 Introduction to MUST

We present MUST (Marmot Umpire Support Tool), a new approach to runtime MPI correctness checking. We designed MUST to overcome the limitations to scalability and extensibility of Umpire and Marmot and their hard coded trace communication with a centralized manager. Its design focuses on the following goals:

1. Correctness
2. Scalability
3. Usability
4. Portability

The correctness goal is the most important one and comes with two sub-goals: first, the tool must not give false positives; second, the tool should detect all MPI related errors that manifest themselves in an execution with MUST. We restrict this second sub-goal to runs in which errors actually occur, as the detection of all potential errors would likely incur an intolerable runtime overhead, which would limit the applicability of the tool.

The second goal, scalability, is one of the main motivations for this new approach to MPI checking. The tool must scale at least to small or medium sized test cases on next generation HPC systems. With the current trend towards high numbers of computing cores, this means at least a range of 1,000 to 10,000 processes. Our goal is to offer a full set of correctness features for 1,000 processes at a runtime overhead of less than 10%, and a restricted set of correctness features for 10,000 processes at the same runtime overhead.

The further goals, usability and portability, are important to achieve a successful tool that will find acceptance with both application programmers and HPC support staff. A common problem with many HPC tools is that they require the application developer to recompile and relink the application, which can be very time consuming for larger applications. Therefore, we aim to avoid this requirement with MUST. Further, tools must be adaptable to special HPC systems that impose restrictions such as no support for threads.

We address both issues with P^nMPI [5], an infrastructure for MPI tools. P^nMPI simplifies MPI tool usage by allowing tools to be added dynamically, removing the need to recompile and offering flexibility in the choice and combination of PMPI-based tools. Only the P^nMPI core is linked to the application, instead of a certain MPI tool. If the MPI tools are available as shared libraries, P^nMPI supports the application of any number of MPI tools simultaneously. Thus, at execution time, the tool user can decide which tools he wants to apply to an application.

P^nMPI achieves this flexibility by virtualizing the MPI Profiling interface. It considers each MPI tool as a module and arranges these modules in stacks that specify the order in which MPI calls are passed to the modules. These modules may also co-operate with each other by offering services to or using services from other modules. Further, special P^nMPI modules allow more enhanced features like condition-based branching in stacks. This infrastructure provides flexibility combined with advantages to tool usability. As a result, we base the design of MUST on P^nMPI and use fine grained modules that can be composed to form an instance of MUST. With this basic infrastructure, we can easily adapt the MUST tool to specialized scenarios such as when only an individual correctness check is of interest.

A further important aspect of MUST is the notion that the overall tool will consist of three layers. The bottom layer is provided by P^nMPI and its modules that provide the basic infrastructure and composability of the tool. The actual correctness checks form the upper layer of MUST. The remaining middle layer has to provide service tasks like trace record generation and the communication of trace records to processes and threads that are exclusively allocated to the tool, which are used to offload correctness analyses. A further task is the management of these processes and threads for error cases, startup and shutdown. This task is tool agnostic and needed for many HPC tools. As a result, we want to provide this layer of functionality as a decoupled set of modules that is also available to other tool developers. Thus, we name this layer of functionality "Generic Tool Infrastructure" (GTI).

5.4 MUST Design

This section introduces some of the key design ideas of MUST. As discussed in the last section, our design uses P^nMPI for the underlying infrastructure along with a set of fine grain modules that implement the MPI checks. A first important aspect of the MUST design is the ability to execute correctness checks either in an application process itself (in the critical path) or in extra processes or threads that are used to offload these analyses (away from the critical path). This choice can provide a low runtime overhead while supporting portability. The first part of this section introduces the concepts that we use to achieve this goal. Afterwards we present an overview of the overall components of the MUST design, and highlight their tasks. A further aspect of the design is the communication of trace records. We present an overview of how different types of modules combine to implement this communication. These modules are part of the GTI layer and can be used by other tools.

5.4.1 Offloading of Checks

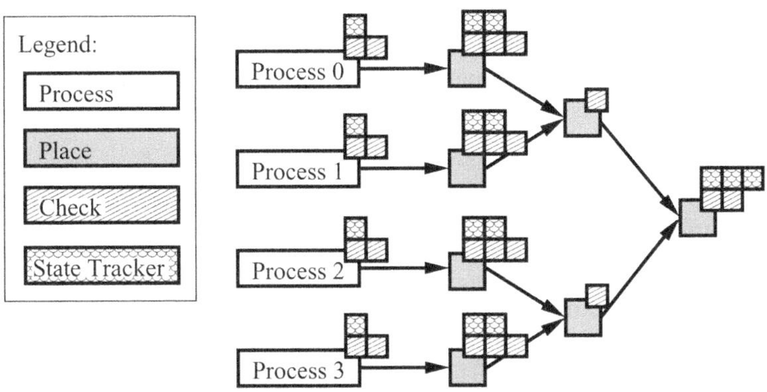

Fig. 5.3: Example instantiation of places, checks, state trackers and a communication network.

The option to execute correctness checks on additional processes or threads is one of MUST's most important aspects. We refer to such a process or thread by the term "place". Marmot and Umpire both execute some checks on an extra place (the *manager* thread for Umpire and the *DebugServer* process for Marmot). However, both tools do not support the selection of the place of execution freely, as these checks are explicitly aware of being executed at a certain place. Moving such a check into the critical path, or a check being executed in the critical path to another place is not easily possible in either Marmot or Umpire.

The main problem is that the execution of checks often requires background information on the state of MPI. It is possible within the application process for a synchronous tool to query such information with MPI calls, while it is not possible on additional processes that do not have access to the MPI library. Similarly, if the MPI process has proceeded beyond the MPI call, the relevant state may have changed. Also, the required information often must be gathered and updated during application execution. For example, determining which requests are currently active requires the sequence of request initiations and completions. While much of the work can be offloaded to MPI emulation, the gathering of the basic information must take place in the application processes themselves.

MUST uses the concept of "state trackers" to solve this problem. All information that a check requires but is not directly available from the arguments of the MPI call that triggered this check, must be provided by state trackers. These trackers are implemented as independent modules and may gather different types of data during the application's runtime and provide it to checks when needed. If multiple checks require the same state tracker, a single instance of the state tracker can provide this information. In order to support the placement of a check at any place, the MUST system has to determine which state trackers are required on each place. This strategy provides a transparent way to implement checks that can be offloaded to places.

Figure 5.3 shows an example distribution of places, checks, state trackers and a trace communication network. It uses four application processes and seven extra places to offload checks. The checks are highlighted as little boxes in the top right corner of the places or application processes. Each place or application process may also need state trackers that are indicated by little boxes above the checks.

5.4.2 Major Components

Figure 5.4 shows the main components of MUST and parts of their overall interaction. The correctness checks and the tool infrastructure are provided as modules from MUST, P^nMPI, and the GTI (top row). We summarize a further set of components in the top right of the figure as "descriptions", which describe properties of some of the modules and formalize what checks apply to the arguments of specific MPI calls. They also characterize the dependencies of checks and state trackers. A GUI (middle left) provides users with options to individualize MUST for their needs, e.g., to specify the checks being used, to add extra processes/threads that offload checks, or to define the layout of the trace communication network. A default configuration will usually by sufficient for smaller test cases, while large scale

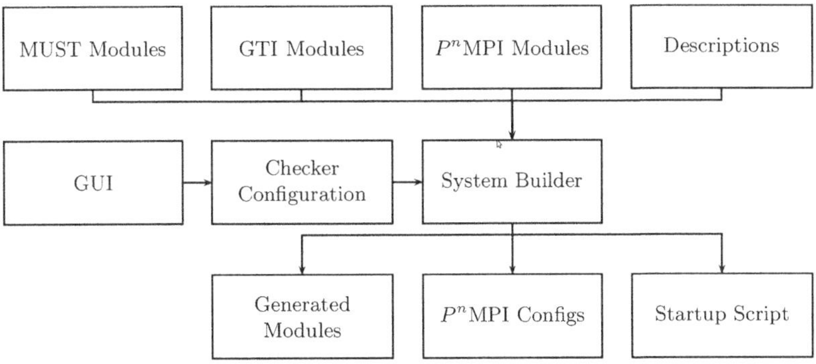

Fig. 5.4: Major components in MUST; arcs denote input/output dependencies.

tests will need a specifically tailored configuration. The system builder component uses the selected tool configuration, the list of available modules and the various descriptions to create the configuration files for P^nMPI and additional intermediate modules, including specialized MPI wrappers to create and forward the necessary trace records. An additional startup script may be provided to simplify the startup of the application with MUST.

5.4.3 *Trace Communication System*

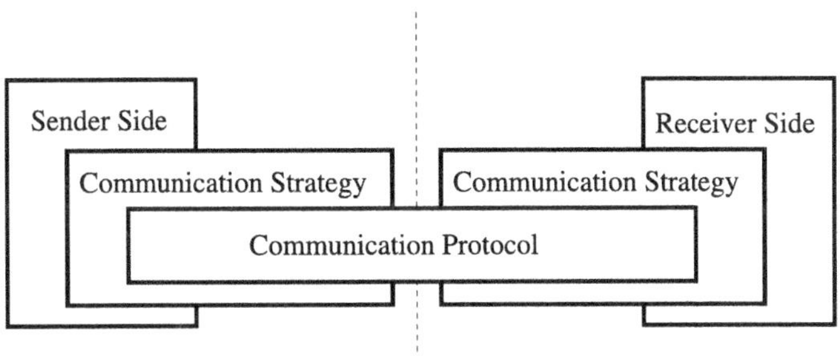

Fig. 5.5: Composition of places with communication strategies and communication protocols.

An important aspect of MUST's design is its encapsulation of how to transport trace records from one process or thread to another. An efficient communication of trace records primarily depends on two things: first, an efficient communication medium that optimizes the use of the underlying system where possible; second, an efficient strategy to use these communication media. Thus, we must use shared memory when communicating on node or rely on InfiniBand instead of Ethernet if both networks are available. It will usually be very inefficient to transfer tiny trace records with single messages with a TCP/InfiniBand/MPI based communication. Also, it will be more effective not to wait until the message has been received for most media.

The GTI component of MUST solves this problem by combining two different types of modules to implement a communication. The first type of module, a "communication strategy", decides when to send what information: it may send trace records immediately or it may delay the transfer of trace records and aggregate them into larger messages. The second type of module, a "communication protocol", implements the communication for a particular communication medium, e.g., TCP/IP, InfiniBand, SHMEM, or MPI.

Figure 5.5 shows how we compose these modules on the sender and receiver side. By selecting appropriate combinations of these two module types, we can provide a flexible, adaptable and high performance communication of MPI communication traces. One instantiation of MUST may use multiple combinations, e.g., a shared memory communication protocol to transfer MPI trace records to an extra thread and a TCP/IP communication protocol to transfer trace records from this thread to a further place used to offload checks.

5.5 Initial Experiments

We developed a proof of concept implementation of a subset of the MUST design in order to verify our ideas, as well as to perform first performance studies. The implementation provides the features necessary to use extra places and transfer trace records to them. One of our early questions is the feasibility of our runtime overhead goals. The question at hand is, whether we can transfer the trace records from the application processes to extra places without perturbing the application. We use initial experiments with our proof of concept implementation to study this problem. We use two different communication layouts and three different communication strategies to study different communication approaches. Our tests intercept all MPI calls and create a trace record for each. We send these trace records from the application processes to extra places and measure the runtime overhead that results from this extra communication. However, the receiver side only receives and unpacks these trace records; no checks are executed. We use NPB3.3-MPI as target applications and run our tests on a 1152 node AMD Opteron Linux cluster with a DDR InfiniBand network. Each node has 8 cores on four sockets and 16 GB of main memory that is shared between all cores.

As the communication protocol we use MPI itself, as it provides an easily available and highly optimized communication medium. It also offers a simple way to allocate extra processes for MUST. We use P^nMPI based virtualization to split an allocated `MPI_COMM_WORLD` into multiple disjoint sets. The application uses one of these sets as its `MPI_COMM_WORLD`, which is transparent to the application itself. The remaining sets can be used for MUST. MPI based communication between all of the sets is possible. We use two different communication layouts, which are "1-to-1", a best case layout where each application process has one extra process that receives its trace records, and "all-to-1", a centralized manager case where all application processes send their trace records to one extra process. The first layout helps to determine what runtime overhead to expect for a case where checks can be well distributed and no centralized manager needs to receive records from all processes. The second case captures the limits of a communication with a centralized manager, as in Umpire and Marmot.

We use three different communication strategies to implement different communication schemes. These are:

Ssend: Sends one message for each trace record, waits for the completion of the receive of each message before it continues execution.

Isend: Sends one message for each trace record, does not wait for the completion of the receive of the message. With the MPI based communication protocol this is implemented with an `MPI_Isend` call.

Asend: Aggregates multiple trace records into one message, sends the message when either the aggregate buffer is full (100KB) or a flush is issued. As with *Isend*, the sender does not wait for the completion of the receive of the message.

The *Ssend* strategy is very similar to the communication currently used in Marmot. Besides its obvious performance disadvantage, it simplifies handling of application crashes as it guarantees that trace records were sent out from the application before a crash can occur. The *Isend* strategy is still simple to implement and should overcome the performance problems of the *Ssend* strategy. The *Asend* strategy, which is similar to Umpire's communication strategy, is our most complex strategy, but offers multiple optimizations that may provide a low runtime overhead. In particular, we expect that this method will achieve higher bandwidth, due to the aggregation of the trace records. However, its performance benefit will depend on a good interleaving of the communication: we expect a high runtime penalty if the aggregated messages are sent while the application is in a communication phase. On the other hand, sending the aggreagated messages while the application is in a computation phase will incur close to no runtime overhead, particularly on systems with communication co-processors. Experiments with Umpire already highlighted the significance of this timing behavior [6]. As a result, we instrument the NPB kernels to issue a flush of the aggregated buffer when the application enters a computation phase. This removes the need for an automatic detection of computation phases and represents a close to best case scenario. For a final system, we will have to apply a heuristic to guess when the application is entering a computation phase.

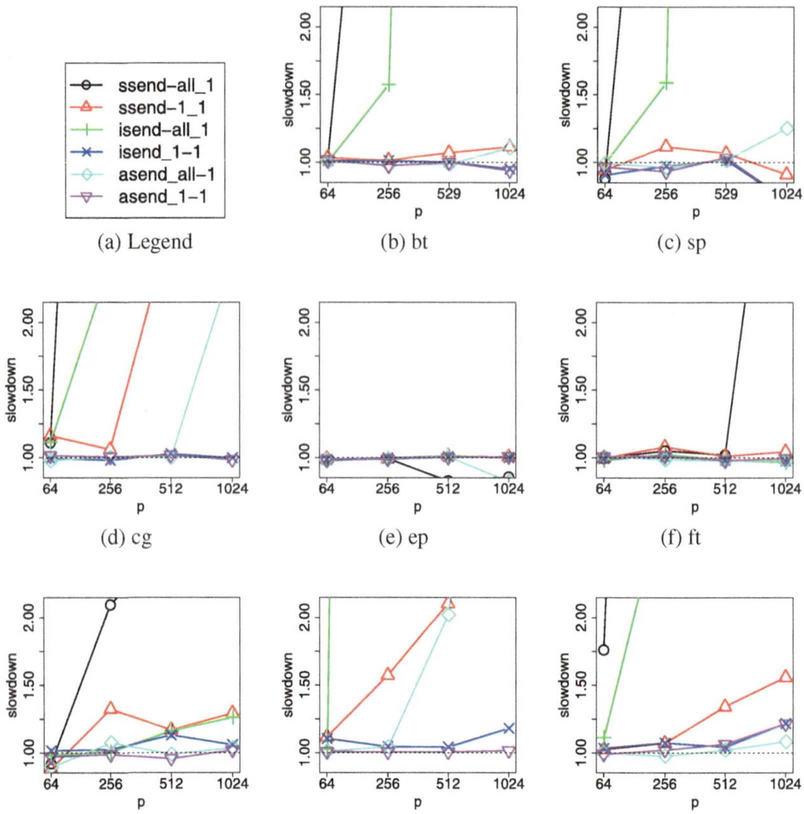

(a) Legend (b) bt (c) sp

(d) cg (e) ep (f) ft

(g) is (h) lu (i) mg

Fig. 5.6: Runtime overhead for different implementations of a trace transfer.

Figure 5.6 summarizes the performance results for NPB3.3-MPI with problem size D when using the three different types of trace communication and both place configurations. Subfigure 5.6a shows the legend for the different communication layout and communication strategy combinations. The remaining figures show the runtime overhead for 64 to 1024 processes for these combinations. The all-to-1 cases fail to scale to 1024 processes for most of the kernels. Where the *Ssend* and *Isend* versions of the all-to-1 communication even fail for 256 processes for most kernels, the *Asend* strategy scales to up to 512 processes. Its main advantage is the reduction in messages arriving at the centralized manager, which leads to a lower workload. For the 1-to-1 cases, the *Ssend* strategy incurs a slowdown of up to 3 and hence fails to meet our performance goals. However, its slowdown does not necessarily increase with scale. Both the *Isend* and *Asend* strategies for the 1-to-1 cases incur a low runtime overhead, even at scale. These strategies only fail to achieve the desired less than 10% runtime overhead for the kernel mg. However, the problem size D of NPB3.3-MPI is a challenging test case at 1024 processes, as the fraction of the total execution time spent in MPI is very high at this scale. We expect better results for most applications.

5.6 Related Work

Several other MPI message checkers exist beyond Marmot [3] and Umpire [4] including MPI-Check [7] and ISP [8]. Both of these tools are not reported to scale to more than a hundred processes. Especially the complex analysis of alternative executions in ISP limits its scalability dramatically. We hope to combine our efforts with the ones for ISP in future work, as both tools have the same basic needs.

The generic tool infrastructure component of MUST relates to a wide range of infrastructure and scalability work. This includes P^nMPI [5], as well as infrastucture tools like MRNet [9], which we may use to implement several of the GTI components. Also, existing HPC tools like VampirServer [10] and Scalasca [11], or debuggers like DDT [12] and Totalview [13] may implement well adapted communication schemes that can be used for the GTI components. Further, these tools, as well as upcoming tools may employ modules of the GTI to implement their communications and may thus benefit from this component.

5.7 Conclusions

This paper presents a novel approach to create a runtime infrastructure for scalable MPI correctness checking. As far as we know, existing approaches – like Marmot and Umpire – lack the scalability needed for large HPC systems. Further, these tools use static communication systems that are hard to adapt to different types of systems. Also the implementation of new checks and the extension of existing ones is hard for these tools, as their checks are tightly coupled to their internal data structures and infrastructures. Our approach overcomes these problems by using a fine-grained module-based design that uses P^nMPI. We present an overview of this design and highlight our most important concepts that allow the offloading of checks to extra processes and threads. Further, we present a flexible communication system that promises an efficient transfer of trace records between different processes or threads. To demonstrate the feasibility of our design and to highlight the performance capabilities of our communication system, we present a performance study with a proof of concept implementation. This study shows that our ambitious runtime overhead goals are feasible, even at scale. In particular we demonstrate full MPI tracing for up to 1024 processes while transferring the trace records to extra processes without perturbing the application.

References

1. Message Passing Interface Forum: MPI: A Message-Passing Interface Standard. http://www.mpi-forum.org/docs/mpi-10.ps (1995)
2. Message Passing Interface Forum: MPI-2: Extensions to the Message-Passing Interface. http://www.mpi-forum.org/docs/mpi-20.ps (1997)
3. Krammer, B., Bidmon, K., Müller, M.S., Resch, M.M.: MARMOT: An MPI Analysis and Checking Tool. In Joubert, G.R., Nagel, W.E., Peters, F.J., Walter, W.V., eds.: PARCO. Volume 13 of Advances in Parallel Computing., Elsevier (2003) 493–500
4. Vetter, J.S., de Supinski, B.R.: Dynamic Software Testing of MPI Applications with Umpire. Supercomputing, ACM/IEEE 2000 Conference (04-10 Nov. 2000) 51–51
5. Schulz, M., de Supinski, B.R.: P^NMPI Tools: A Whole Lot Greater Than the Sum of Their Parts. In: Supercomputing 2007 (SC'07). (2007)
6. Hilbrich, T., de Supinski, B.R., Schulz, M., Müller, M.S.: A Graph Based Approach for MPI Deadlock Detection. In: ICS '09: Proceedings of the 23rd international conference on Supercomputing, New York, NY, USA, ACM (2009) 296–305
7. Luecke, G.R., Zou, Y., Coyle, J., Hoekstra, J., Kraeva, M.: Deadlock Detection in MPI Programs. Concurrency and Computation: Practice and Experience **14**(11) (2002) 911–932
8. Vakkalanka, S.S., Sharma, S., Gopalakrishnan, G., Kirby, R.M.: ISP: A Tool for Model Checking MPI Programs. In: PPoPP '08: Proceedings of the 13th ACM SIGPLAN Symposium on Principles and practice of parallel programming, New York, NY, USA, ACM (2008) 285–286
9. Roth, P.C., Arnold, D.C., Miller, B.P.: MRNet: A Software-Based Multicast/Reduction Network for Scalable Tools. In: SC '03: Proceedings of the 2003 ACM/IEEE conference on Supercomputing, Washington, DC, USA, IEEE Computer Society (2003) 21
10. Brunst, H., Kranzlmüller, D., Nagel, W.E.: Tools for Scalable Parallel Program Analysis - Vampir NG and DeWiz. The International Series in Engineering and Computer Science, Distributed and Parallel Systems **777** (2005) 92–102
11. Wolf, F., Wylie, B., Abraham, E., Becker, D., Frings, W., Fuerlinger, K., Geimer, M., Hermanns, M., Mohr, B., Moore, S., Szebenyi, Z.: Usage of the SCALASCA Toolset for Scalable Performance Analysis of Large-Scale Parallel Applications. In: Proceedings of the 2nd HLRS Parallel Tools Workshop, Stuttgart, Germany (July 2008)
12. Edwards, D.J., Minsky, M.L.: Recent Improvements in DDT. Technical report, Alinea, Cambridge, MA, USA (1963)
13. Totalview Technologies: Totalview - Parallel and Thread Debugger. http://www.totalviewtech.com/products/totalview.html (July 2009)

Chapter 6
HPC Profiling with the Sun Studio™ Performance Tools

Marty Itzkowitz and Yukon Maruyama

Abstract In this paper, we describe how to use the Sun Studio Performance Tools to understand the nature and causes of application performance problems. We first explore CPU and memory performance problems for single-threaded applications, giving some simple examples. Then, we discuss multi-threaded performance issues, such as locking and false-sharing of cache lines, in each case showing how the tools can help. We go on to describe OpenMP applications and the support for them in the performance tools. Then we discuss MPI applications, and the techniques used to profile them. Finally, we present our conclusions.

6.1 Introduction

High-performance computing (HPC) is all about performance. This paper describes the various techniques implemented in the Sun Studio Performance Tools to profile HPC applications. We first describe how users can recognize the symptoms of performance problems. We then discuss problems common to single-threaded programs, and then go on to describe additional issues that manifest in multi-threaded programs. We then describe the characteristics of two of the main HPC programming models, OpenMP and MPI, and review the specific performance issues with each, and show how the tools can help.

6.1.1 The Sun Studio Performance Tools

The Sun Studio Performance Tools are designed to collect performance data on fully optimized and parallelized applications written in C, C++, Fortran, or Java,

Marty Itzkowitz, Yukon Maruyama
Sun Microsystems, 16 Network Circle, Menlo Park, CA 94025, USA
e-mail: {marty.itzkowitz, yukon.maruyama}@oracle.com

M.S. Müller et al. (eds.), *Tools for High Performance Computing 2009*,
DOI 10.1007/978-3-642-11261-4_6, © Springer-Verlag Berlin Heidelberg 2010

and any combination of these languages. Data is presented in the context of the user's programming model. With appropriate settings, measurements can be done at production-scale, in terms of the numbers of threads and processes, the size of the address-space, and length of run.

The tools support code compiled with the Sun Studio or GNU compilers. They also work on code generated by other compilers, as long as those compilers produce compatible standard ELF and DWARF symbolic information.

The tools run on the Solaris™ or Linux operating systems, on either SPARC® or x86/x64 processors. The current version, Sun Studio 12 update 1, is available for free download [1], and was used to record the screen-shots in this paper.

The objective in designing the tools was to minimize the number of mouse clicks that it takes to reach the point at which the performance problem is shown.

6.1.2 The Sun Studio Performance Tools Usage Model

The usage model for the performance tools consists of three steps. First, the user compiles the target code. No special compilation is needed, and full optimization and parallelization can be used. It is recommended that the **-g** flag be used to get symbolic and line-number information into the executable. (With the Sun Studio compilers, the **-g** flag does not appreciably change the generated code.)

The second step is to collect the data. The simplest way to do so is to prepend the **collect** command with its options to the command to run the application. The result of running **collect** is an experiment which contains the measured performance data. With appropriate options, the data collection process has minimum dilation and distortion, typically about 5%, but significantly larger for tracing runs.

The third step in the user model is to examine the data. Both a command-line program, **er_print,** and a GUI interface, **analyzer**, can be used to examine the data. (The "**er**" refers to "experiment record," the original name for what is now called an experiment.) Much of the complexity introduced into the execution model of the code comes from optimizations and transformations performed by the compiler. The Sun compilers insert significant compiler commentary into the compiled code. The performance tools show the commentary, allowing users to understand exactly what transformations were done.

6.1.3 The Sun Studio Performance Tools Features

The Sun Studio performance tools support data collection by statistical sampling of callstacks, based on either clock-ticks or hardware-counter overflow events. They also support data collection based on tracing synchronization API calls, memory allocation and deallocation API calls, and MPI API calls.

They can show a list of functions, annotated by metrics, both exclusive (in the function itself) or inclusive (in the function, and any that it calls). They also support a caller-callee view, and annotated source or disassembly listings. They can show a list of source lines or instructions, annotated with metrics, and a graphical timeline showing profiling events as a function of time. Other views are specific to various programming models or data collected and will be described below.

All of the data can be filtered in various ways to drill down into specific performance problems in function, or source-lines, or calling context, as well as by thread or CPU.

6.1.4 Diagnosing Performance Problems

The first step in understanding performance problems is determining whether or not a problem exists. That understanding is best achieved with a repeatable example (benchmark) that uses input data and problem size of a scale comparable to the production runs for which the program is being tuned. From such runs, many techniques can be used to determine if there is a problem.

Many problems have an intrinsic scale factor, N, and typical high-performance computing codes are intended to run at the largest practical scale. Performing measurements for different values of N can show whether the performance of the application scales with N, or $ln\ N$, or N^2 or even a higher power of N. If simple tests show that there are scaling problems with the application, more detailed data, including clock- and hardware-counter statistical profiling data and various kinds of tracing data, can be collected to isolate and fix the problem.

The most important question to ask is "what can I change to improve the performance of the application"? To answer that question, data is presented in the context of the user model, showing what resources are being used by the application, and where in the application they are being used. The data can also show how the execution got to that point in the program.

The next section of this paper describes various single-threaded application performance issues; the following section explores additional issues presented by multi-threaded applications. The fourth section discusses issues relating to the OpenMP programming model, and the fifth section describes issues relating to MPI programs. In the last section, we present our conclusions.

6.2 Single-Threaded Application Performance Issues

The main issues for single-threaded applications are CPU algorithmic inefficiency and memory subsystem performance.

Two techniques are used in the tools to explore these issues: clock-based profiling, used on various UNIX systems for at least 25 years [2]; and hardware-counter profiling first described in 1996 [3].

6.2.1 Algorithmic Inefficiency

The signature of algorithmic inefficiency is high resource consumption for parts of the program that do not represent the actual computational core of the algorithm. We will use two examples of low-hanging fruit to show how the tools can be used to find such signatures.

The most straightforward way to look for algorithmic inefficiency is to use clock-based statistical callstack sampling, the default experiment recorded with the Sun Studio Performance tools. On the Solaris operating environment, clock profiling collects metrics of User CPU Time, System CPU Time, Wait CPU Time, User Lock Time, Data Page Fault Time, Text Page Fault Time, and Other Wait Time. On Linux, only User CPU Time can be monitored. Such measurements show where the resources are being consumed during execution, but do not tell you whether that is bad or good. The user must decide whether the resource-consumption is necessary for the computation, or if it can be optimized.

The example program used is very simple, with two operations, each coded in an efficient way, and in an inefficient way. The two pairs of functions are named **good_init/bad_init**, and **good_insert/bad_insert**. The first of these is an apparently trivial botch which consists of doing a static initialization many times within a loop, instead of once before the loop. Figure 6.1 shows the function list from a clock-profiling experiment on the program.

The two functions highlighted are the two versions of initialization. Although the Exclusive CPU Time (column 1) is about the same, the Inclusive CPU Time (column 2) is very different, reflecting the time spent in **static_init_routine**. Figure 6.2 show the source for both of the initialization functions, clearly showing the high inclusive CPU usage in the **bad_init** version.

While this may seem too obvious for anyone ever to program, similar problems may arise more subtly, especially if the programmer is using APIs written by others.

This example is based on a performance problem in a commercial parallelization tool product developed by one of us (MI). The user interface of the product presented a list of loops in a program, with an icon representing the parallelization state of the loop. The application was written to a library that provided two APIs to add icons to a table. One added an icon to a specific row, and was easy to code; the second added a vector of icons to the table. When used on targets with relatively few loops, no problem was seen, but when used on a target with more than 900 loops, the interface took more than 20 minutes to show up. The root cause was that using the simpler algorithm caused a recomputation of the table geometry with each icon added, while the vector API only recomputed the geometry once. It took only 23 seconds to come up, a 60X improvement.

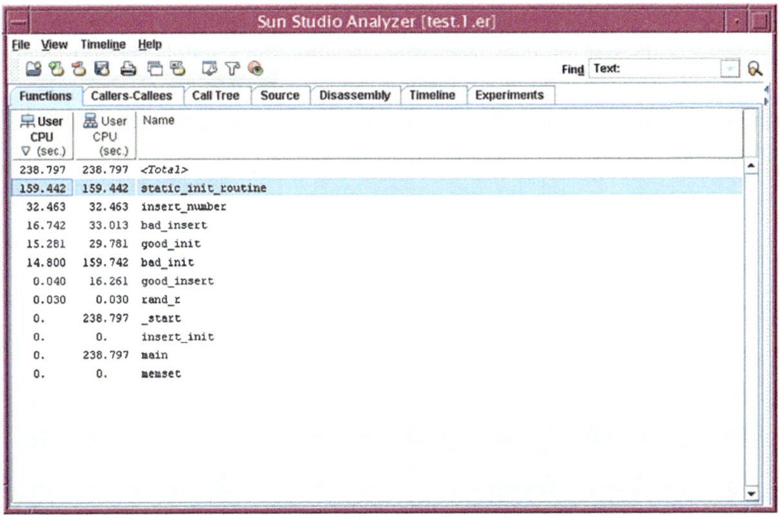

Fig. 6.1: Function List from the **lowfruit.c** program

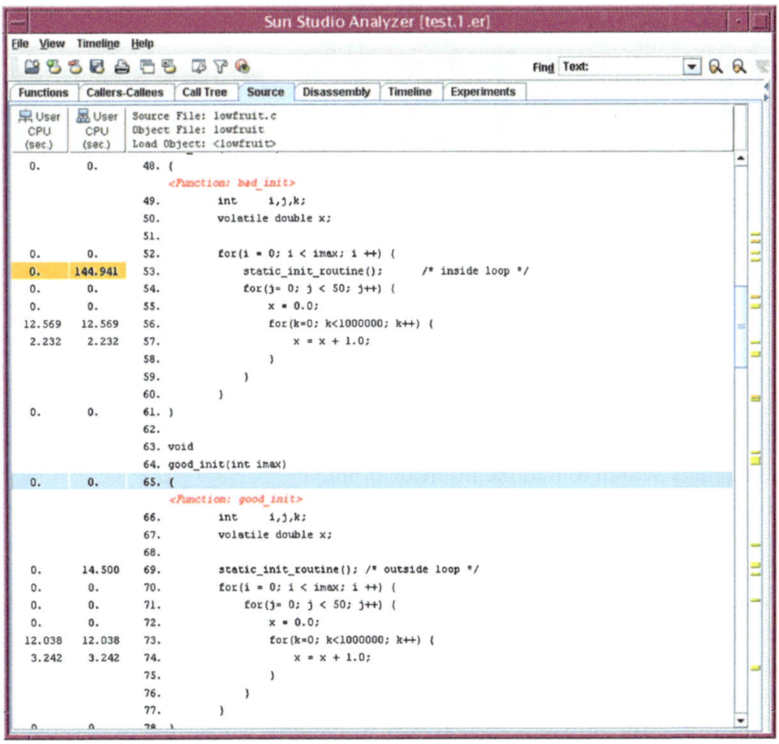

Fig. 6.2: Source Display for **good_init** and **bad_init**

The second example is a typical one where at small scale, no problem is manifest, but at large scale performance drops dramatically. The example shows two different ways to insert an element in an ordered list. Figure 6.3 shows the source of the inefficient version, **bad_insert**. It does a linear search to determine where to insert the next element.

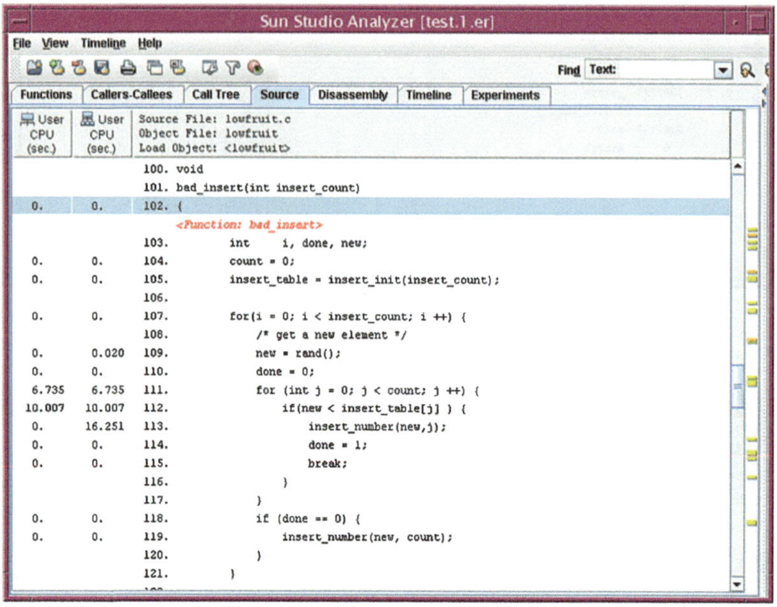

Fig. 6.3: Source Display for **bad_insert**

Note the approximately 17 seconds spent on lines 111 and 112, which represents the time spent determining where to insert the next element.

By contrast, Figure 6.4 shows the efficient version of the same problem, which does a binary search to find the insertion point.

Note that less than 0.1 seconds is spent determining where to insert the next element. The time to perform the actual insertion is approximately 16 seconds and is the same in both versions.

The above examples show performance issues in the user source code. For some problems, the issues arise in the compiler's code-generation. In those cases, the disassembly of the code, with per-instruction performance data, can be used to understand the behavior.

Clock profiling provides a good way to identify where the users should focus their tuning efforts. Sometimes, the problem areas uncovered by clock profiling can not be fully explained by algorithmic inefficiency; often, the user needs to also de-

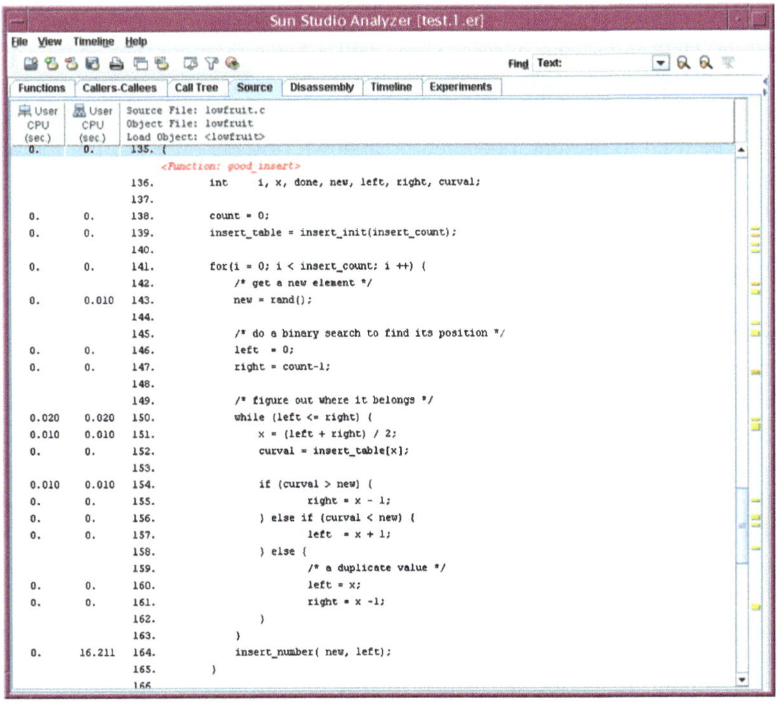

Fig. 6.4: Source Display for **good_insert**

termine how the program is using the underlying hardware, especially the memory subsystem.

6.2.2 Memory Subsystem Performance Issues

In modern computer systems, access to memory is mediated by various components in the memory subsystem. It contains hardware that maps virtual addresses to physical memory pages (a translation-lookaside-buffer, or TLB) and one or more levels of cache, designed to minimize the latency of memory fetches. These elements of the memory subsystem are not directly in the user-model of the computation. The efforts of the hardware designers to minimize latency make it difficult to relate the performance issues directly to the code in which they occur.

To help understand how a program interacts with the hardware, most modern chips have counters to measure the performance of the CPU and other subsystems in the machine. The counters can be used either by reading them directly, or by statistical profiling based on counter overflows. While the support for counters varies from chip to chip, most chips have counters that measure memory-related activity, including TLB and cache misses. Some CPUs have counters that count not only cache-miss events, but also measure the cost of accessing memory, for example, the number of cycles stalled waiting for a cache-miss to be satisfied. Hardware counters can be used to present a detailed breakdown of CPU time waiting for various components of the memory subsystem.

Operating-system support for hardware-counter profiling is included in Solaris in **libcpc.so**. Linux systems require kernel support and a user API, such as **perfctr** [4], **perfmon2** [5], or **PCL** [6], to enable hardware-counter profiling. The Sun Studio Performance Tools currently support only the **perfctr** API on Linux systems. Users can collect hardware-counter overflow profiles based on whatever counters are available on the particular chip being utilized. Invoking the **collect** command with no arguments will print a list of all the counters available on that system.

As an example of the use of hardware-counter profiles, we constructed a simple program, **cachetest**, that does an identical matrix-vector-multiply computation in eight different ways. This code was originally intended to demonstrate the effects of optimization, so the source code consists of four copies of the same source file (with different function names), compiled with different optimization levels: no optimization, **-O** optimization, **-fast** optimization, and auto-parallelization. One version of the computation in each file is in row-column order and other is in column-row order. The eight functions are named **dgemv_***, with a suffix indicating the optimization level (**g**, **opt**, **hi**, or **p**) and the loop-order (either **1** or **2**).

The data were recorded on a Solaris 10 system with an UltraSPARC® III-Cu processor. That chip has a TLB and two levels of cache, a first-level cache called the D-cache, and a second-level cache called the E-cache.

Two experiments were recorded, one collecting clock profiles and hardware-counter profiles for cycles and combined D-cache & E-cache stall cycles, and the second collecting E-cache stall cycles and I-cache stall cycles. The two experiments are merged, and Figure 6.5 shows the function list, sorted alphabetically.

There are a number of points of interest in this data.

First note that no appreciable time is spent dealing with instruction-cache stalls (column 5). The program is too small for I-cache performance to be an issue.

Next, note that for some functions, User CPU Time (column 1, based on clock-profiling) and CPU Cycle Time (column 2, based on hardware counter profiling for cycles) are different. Although one might naively expect them to be the same, they are not. User CPU Time represents the time the operating system thinks the process was running in user mode. CPU Cycle Time represents the time the chip thinks the process was running in user mode. They differ in an important way: in processing a TLB miss, the operating system does not change its notion of whether the process is running in user-mode – to do so would significantly increase the overhead of processing the miss. Thus the difference between these two metrics represents the time lost due to TLB misses. (There are other events that can contribute to the difference, but they are not relevant to this program.)

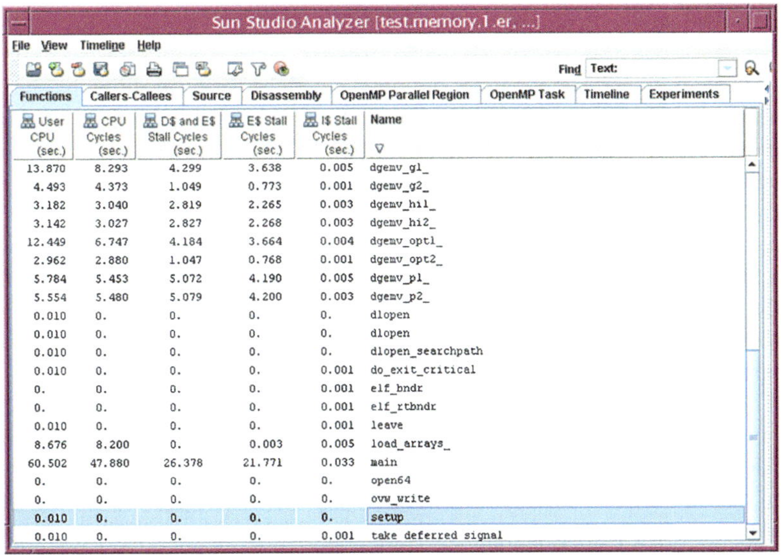

Fig. 6.5: Function List from **cachetest**

Optimization affects both the memory performance and the efficiency of the actual generated code. From the numbers shown, the breakdown of time spent is straightforward. For the slowest version, **dgemv_g1**, a total CPU time of ~13.9 seconds is broken down into ~5.6 seconds lost due to TLB misses, ~3.5 seconds lost due to E-cache misses, ~0.7 seconds lost due to D-cache misses, and ~4.0 seconds of real computation. The fastest version, **dgemv_opt2**, shows little time lost due to TLB misses, ~0.7 seconds lost due to E-cache misses, ~0.3 seconds lost due to D-cache misses, and ~2.0 seconds of real computation.

Also note that at the lowest optimization levels, **g** and **opt,** the **1** versions are significantly slower than **2** versions. The loop-orders stride through memory differently: one is relatively efficient in cache utilization, while the other one is not. This difference disappears in the **hi** and **p** versions, because the compiler understands the stride-order implications for cache performance, and at high optimization, it interchanges the order of the two loops, so that **1** versions use the same efficient order as the **2** versions. The Sun Studio compilers insert commentary explaining the interchange into the object code, and the commentary is displayed with the source code in the Analyzer.

Hardware counter profiling can also be used to understand other performance aspects of machine behavior (branch misprediction, microcode assists, *etc.*). Tuning at this level requires an intimate understanding of the CPU architecture, and we will not discuss these issues further.

6.2.2.1 Dataspace Profiling

An extension to hardware counter profiling called dataspace profiling has been implemented [7] to better understand the data that is responsible for the cache misses. The data collector attempts to backtrack from the interrupt PC to find the actual instruction causing the cache event. From that instruction and the registers at the time of the interrupt, the collector can usually construct the virtual address being referenced. It then asks the operating system for the corresponding physical address. The Sun Studio compiler outputs information associating each load and store instruction to the symbol table entries of the data being referenced. With these pieces of information, the tools can show cache misses *vs.* the data structures and elements in the program.

The Sun Studio Performance Tools can perform dataspace profiling on SPARC®/ Solaris systems, but not on other chips or operating systems. The reason is that on SPARC® processors, we can backtrack in address space to find the causal instruction, but on x86/x64, that backtracking can not be done. With the advent of new hardware and operating system mechanisms for instruction-sampling, precise instruction and virtual and physical data addresses may be directly captured on both SPARC® and x86/x64 systems.

The work described in [7] was done on one of the SPEC CPU2000 benchmarks, **`mcf`**. By using dataspace profiling, and presenting a display of time-lost due to cache misses against the data structures in the program, and the fields within them, sufficient insight was obtained to yield a 20% decrease in run time for this real application.

6.3 Multi-threading Performance Issues

Multi-threaded applications have the same potential performance issues as single-threaded applications with some additional issues relating the thread interactions and competition for machine resources.

Two of the most important issues are lock contention and false-sharing of cache lines. They are discussed in the next two subsections.

6.3.1 Lock Contention

To ensure consistency of shared data structures in an application, updates to the structures must be done atomically. Atomicity is often implemented using locks—the application must ensure that no thread alters a shared data structure without acquiring a lock governing that structure. If the application has many threads trying to update a structure, contention for the lock can lead to significant application slowdown.

One of the most important factors in the choice of locking strategies is the scope of a lock. Program safety is most easily guaranteed by using locks covering large amounts of data; program efficiency, specifically the minimization of lock contention, is most easily guaranteed by using locks covering relatively small amounts of data.

The Sun Studio Performance tools support measurement of lock contention with a technique called synchronization tracing. During data collection, the measurement library interposes on the standard functions for managing mutex-locks, reader-writer-locks, *etc*. By wrapping these calls, the collector can determine how much time was spent waiting to acquire a lock, and report that as a metric. To minimize the volume of data collected, only those synchronization events that take longer than some threshold are recorded. The threshold can be specified, or will default to approximately five times the calibrated time to acquire an uncontended lock.

Figure 6.6 shows the function list from a test program, **mttest**. The program queues up a number of work blocks, and then spawns a number of threads. Each thread fetches a work block from the queue, synchronizes with the other threads based on a parameter set in the work block, and then does a computation. Two functions, **lock_global** and **lock_local**, represent the same work done with two different synchronization methods. As the names would imply, **lock_global** uses a global mutex so that the threads can only run one at a time, while **lock_local** uses a mutex that is local to the work block, and therefore does not have any contention. In this example, the code was run with four work blocks and four threads.

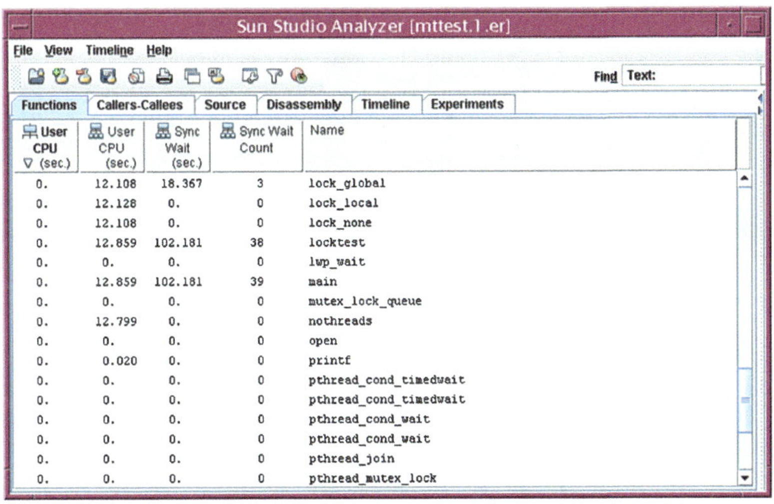

Fig. 6.6: Function List from **mttest**

The two functions, **lock_global** and **lock_local** each consume ~12 seconds of CPU time, reflecting the fact that the processing of the work block is independent of synchronization. However, **lock_global** shows ~18 seconds spent waiting for the lock, while **lock_local** spends no time.

With the global lock, one thread immediately acquires the lock and does 3 seconds of computation while the other three threads wait. After the first thread completes, a second thread acquires the lock and does 3 seconds of computation while the two other threads wait. When the second completes, the third acquires the lock and computes while the last thread waits. Finally, the last thread acquires the lock and does its computation.

The total wait time is 3 X 3, plus 3 X 2, plus 3 X 1, adding up to the 18 seconds shown as synchronization wait time. Each of the three threads that wait also contributes one synchronization wait event to the count.

Figure 6.7 shows the source of **lock_global**, with the synchronization wait on the source line that calls the lock function.

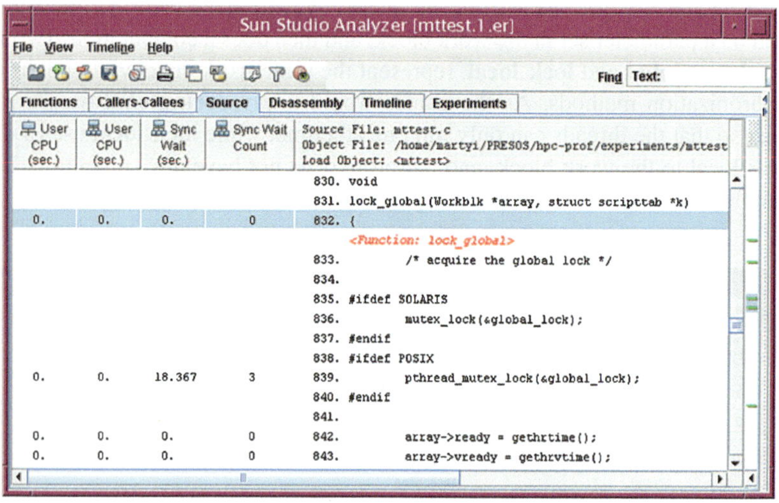

Fig. 6.7: Source of **lock_global**

In this example, it is clear that there is no need to have a global lock; in typical programs, the appropriate scope of a lock is harder to determine. In more complex cases, synchronization tracing is a valuable technique for determining which locks should be targeted for optimization efforts.

6.3.2 False Sharing of Cache Lines

Thread interactions around memory caches are another important performance issue in multi-threaded applications. Multiprocessors have hardware mechanisms to ensure that memory modification by one CPU will invalidate copies of the corresponding cache-line on other CPUs. When one thread updates the data, the cache lines in the CPUs running the other threads will be evicted, and the next reference will force a cache miss and memory access.

If the two threads are referring to the same data, the performance costs of the repeated memory accesses are unavoidable. However, if the two threads are referring to different data, but the data is on the same cache line, the repeated evictions and memory fetches are not really needed, but the hardware can't tell. That circumstance is called "false sharing."

With dataspace profiling, each memory-counter profiling event contains the virtual and physical addresses being referenced. With knowledge of the cache mapping algorithm, an event can be mapped to a cache line.

The performance tools have very powerful filtering mechanisms, and false sharing can be detected using those mechanisms. In the **mttest** example, one of the compute functions exhibits false sharing. Dataspace profile data shows that there is only one hot cache line. By filtering to show only data referring to that cache line, we can determine that it is referred to by four threads, each of which is using a different address within that line. The techniques have been described in detail [8].

6.4 OpenMP Performance Issues

A key challenge for a user-friendly profiling tool is relating information gathered from low-level machine instructions to the user's source code which is written in a high-level programming model like OpenMP. OpenMP programs have a simple fork-join model that is governed by directives in the source code. In the user model, when a parallel region is entered, additional worker threads are created, and, when the computations inside the region are completed, the worker threads disappear. OpenMP 3.0 introduces an additional model, tasking, whereby threads queue up tasks to be performed and worker threads pick up and perform those tasks.

The underlying execution model for OpenMP is significantly more complex than the user model. One of the challenges of any profiling tool is to figure out how to represent the execution-model data back in the user model. Figure 6.8 shows the user-model and execution-model callstacks when the program is executing within a parallel region.

All four threads in this example are executing in the same parallel region, although the leaf PC is shown on different lines in different threads. In the execution model, the function **foo** calls into the OpenMP runtime which dispatches work to the three slave threads as well as back to the master thread. The function called to

Master	Slave1	Slave2	Slave3
foo, line nn main _start	foo, line mm main _start	foo,line pp main _start	foo, line rr main _start

(a) User-model Callstack

Master	Slave1	Slave2	Slave3
foo-OMP, line nn libmtsk foo main _ start	foo-OMP, line mm libmtsk lwp_start	foo-OMP, line pp libmtsk lwp_start	foo-OMP, line rr libmtsk lwp_start

(b) Execution-model Callstack

Fig. 6.8: User-model and Execution-model Callstacks

do the work is shown as **foo-OMP**; it is a so-called outline function constructed by the compiler, but not part of the user model. The line numbers refer to the original source file.

The user-model callstacks are constructed by stripping the frames below the OpenMP runtime in the execution-model in the master, stripping the frames above the OpenMP runtime in the slaves and the master, and stitching the two pieces together to yield the user-model callstacks [9].

When profiling OpenMP applications, the collector records data representing the OpenMP runtime's notion of what the application is doing with every profiling tick. In the current tools, two metrics are computed: OMP Work Time and OMP Wait Time. OMP Work Time includes both serial and parallel CPU time. OMP Wait Time includes overhead as measured in the runtime, and implicit or explicit wait. On the Solaris operating system, OMP Wait Time accumulates whether specified as a busy-wait, consuming CPU time, or as a sleep-wait, not consuming CPU time. On Linux, it accumulates only with busy-wait.

There are four major issues in understanding the performance of OpenMP programs. They are excess parallel overhead, insufficient parallelism, lock contention and synchronization, and load imbalance. They will be discussed in the next four subsections.

6.4.1 Excess Parallel Overhead

Excess parallel overhead arises from applications which are parallelized, but where the work to set up the parallelism is a significant fraction of the total work done in parallel. It is difficult to directly measure: while the OpenMP Performance Measurement API does report time spent in an overhead state, additional work is done in the application to prepare for parallel execution, and that overhead is not detected. (Future work is directed to a more accurate and explicit measurement.)

In the current version of the tools, a pseudo-function, **<OMP-overhead>**, gets metrics attributed to it whenever the program has its leaf function inside the OpenMP runtime. That function is shown as called from the particular parallel regions or tasks responsible for the overhead, allowing the user to see how the program got to that point.

6.4.2 Insufficient Parallelism

Too little parallelism is manifested by high CPU time spent in non-parallel code. According to Amdahl's law, the amount of serial work limits scalability of the application, and thus should be minimized to extract the maximum performance and scalability out of the machine.

In the OpenMP model, all programs start in what is called the "Implicit OpenMP Parallel Region." All serial code is executed in that region, despite its being called a parallel region, while parallel code is executed in other parallel regions. By filtering the performance data to show only data relating to the Implicit OpenMP Parallel Region, direct measurement of serial execution is shown.

Figure 6.9 shows the Parallel Region Tab of an application, with the Implicit OpenMP Parallel Region selected and used to set a filter.

Figure 6.10 shows the function list with the filter applied; it thus shows only the serial portions of the computation.

The function named **serial_** contains an expensive loop that is being executed in serial mode, and represents an opportunity for improving parallelism.

The above example represents the simplest case of not having sufficient parallelism in the code. There are many other cases where this might be a problem. For example, a code may be parallelized using OpenMP sections, where the user has specified 32 threads, but only coded 4 sections. In that case, 28 threads will be doing nothing, despite the code being parallelized. The discrepancy will show up as a load balance issue (see section 6.4.4). Another cause of insufficient parallelism may be in queueing and processing tasks, a topic for future discussion.

6.4.3 Lock Contention

Lock contention causes one or more threads to wait for execution on a lock. It is easily detected as high OMP Wait Time that shows up on the statement representing the lock. It can occur either as an lock call, or as an OpenMP "critical" or "atomic" directive or pragma.

Figure 6.11 shows a picture of source from a program that has a performance hit representing contention for a critical region.

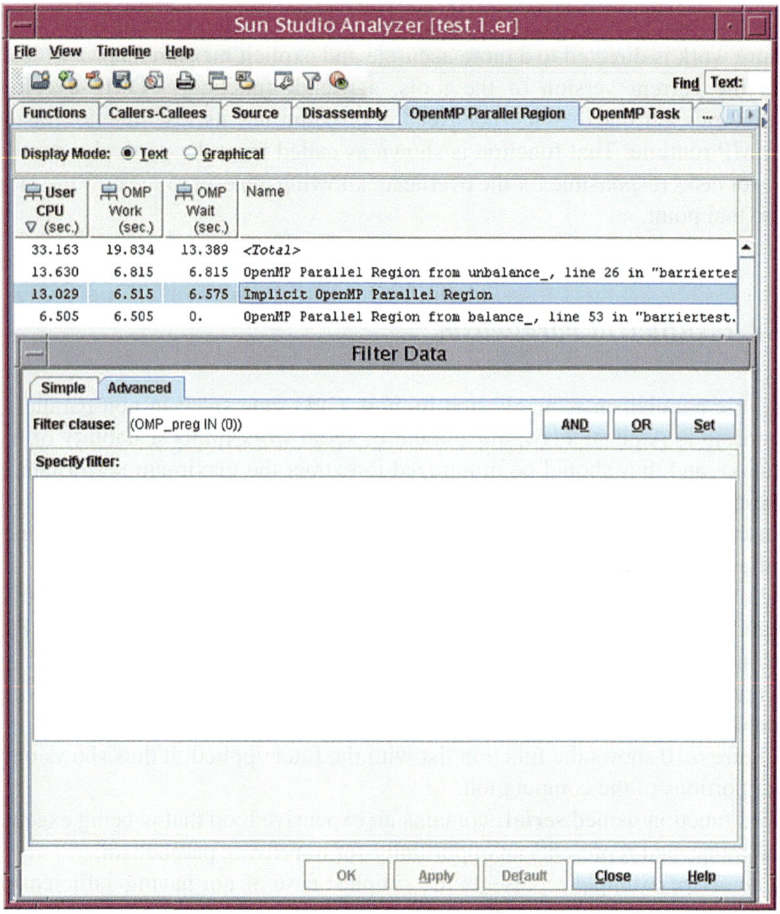

Fig. 6.9: Parallel Region Tab and Filter

The lines with high metric values are highlighted, pointing out the contention. Other scenarios will also show high OMP Wait Time in other pragmas.

6.4.4 Load Imbalance

Load imbalance also shows up as high OMP Wait Time. At the end of a parallel region, synchronization creates an implicit barrier, and the time spent at the barrier represents load imbalance: some threads are done, but none can proceed until all threads reach the barrier. Time spent in that synchronization is attributed to the artificial function <**OMP-implicit_barrier**>.

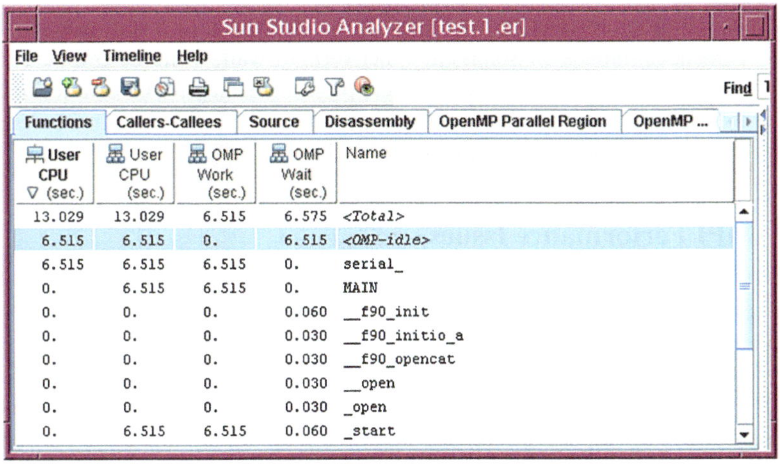

Fig. 6.10: Function list, Filtered to Show only Serial Portions of the Code

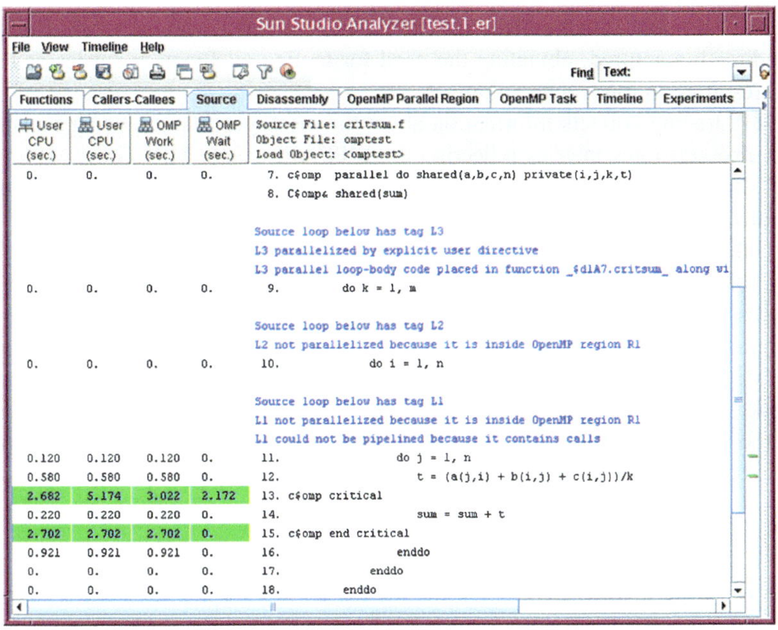

Fig. 6.11: High OMP Wait Time in a Critical Section

OpenMP programs can be run either with a sleep-wait or a spin-wait. OpenMP Wait Time accumulates on Solaris in either form of wait, while CPU time accumulates only for a spin-wait. (On Linux, where profiling is only for CPU time, OpenMP Wait Time is also only accumulated for spin-waits.)

6.5 MPI Performance Issues

MPI programs run as a number of distinct processes, on the same or different nodes of a cluster. Each process does part of the computation, and the processes communicate with each other by sending messages.

The challenge in parallelizing a job with MPI is to decide how the work will be partitioned among the processes, and how much communication between the processes is needed to coordinate the solution. To address these aspects of MPI performance, data is needed on the overall application performance, as well as on specific MPI calls.

Communication issues in MPI programs are explicitly addressed by tracing the application's calls to the MPI runtime API. The data is collected using the VampirTrace [10] hooks, augmented with callstacks associated with each call. Callstacks are directly captured, obviating the need for tracing all function entries and exits, and resulting in lower data volume.

MPI tracing collects information about the messages that are being transmitted and also generates metrics reflecting the MPI API usage: MPI Time, MPI Sends, MPI Receives, MPI Bytes Sent and MPI Bytes Received. Those metrics are attributed to the functions in the callstack of each event.

Unlike many other MPI performance tools, the Sun Studio Performance Tools can collect statistical profiling data and MPI trace data simultaneously on all the processes that comprise the MPI job. In addition, during clock-profiling on MPI programs, state information about the MPI runtime is collected indicating whether the MPI runtime is working or waiting. State data is translated into metrics for MPI Work Time and MPI Wait Time. State data is available only with the Sun HPC ClusterTools™ 8.1 (or later) version of MPI, but trace and profile data can be captured from other versions of MPI.

6.5.1 Computation Issues in MPI Programs

The computation portion of an MPI application may be single-threaded or multithreaded, either explicitly or using OpenMP. The Sun Studio Performance Tools can analyze data from the MPI processes using any of the techniques described in the previous sections for single- and multi-threaded profiles. The data is shown aggregated over all processes, although filtering can be used to show any subset of the processes. Computation costs are shown as User CPU Time (with clock-profiling); computation costs directly attributable to the MPI communication are shown as MPI Work time, a subset of User CPU Time. Time spent in MPI is shown as MPI Time, which represents the wall-clock time, as opposed to the CPU Time, spent within each MPI call.

The data is shown in the function list, Figure 6.12.

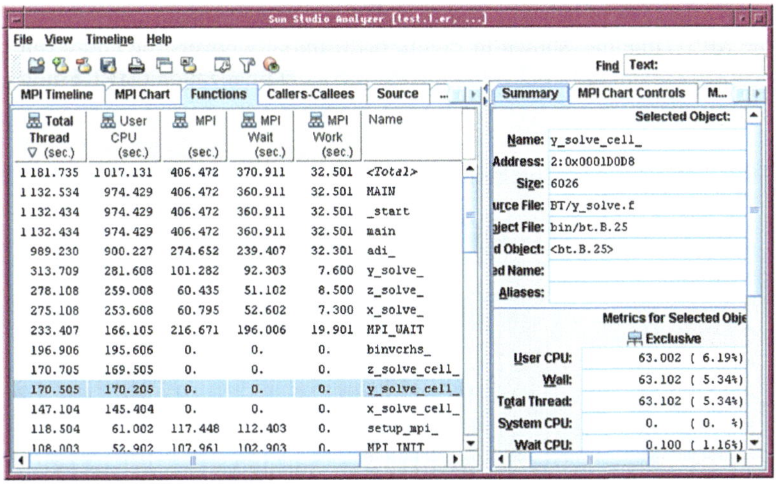

Fig. 6.12: MPI Function List

Functions, such as **y_solve_cell_**, that have high User CPU Time but little or no MPI Work Time or MPI Wait Time represent the actual computations that are done. All of the techniques discussed earlier are relevant to understanding the performance of the computational part of the application, and the tuning that would be done for them is exactly analogous to what would be done for a non-MPI program.

6.5.2 Parallelization Issues in MPI Programs

While the performance issues in computation can be recognized using the techniques described above, problems in partitioning and MPI communication can be recognized by excessive time spent in MPI Functions. The causes of too much time in MPI functions may include: load imbalance; excessive synchronization; computation granularity that is too fine; late posting of MPI requests; and limitations of the MPI implementation and communication hardware.

Many MPI programs are iterative in nature, either iterating on a solution until numerical stability is reached, or iterating over time steps in a simulation. Typically, each iteration in the computation consists of a data receive phase, a computation phase, and a data send phase reporting the results of the computation.

6.5.2.1 Using the MPI Timeline to Visualize Job Behavior

The MPI Timeline gives a broad view of the application behavior, and can be used to identify patterns of behavior and to isolate a region of interest.

The MPI Timeline, shown in Figure 6.13, initially covers the entire run of the application, including initialization (**MPI_Init**) and finalization (**MPI_Finalize**).

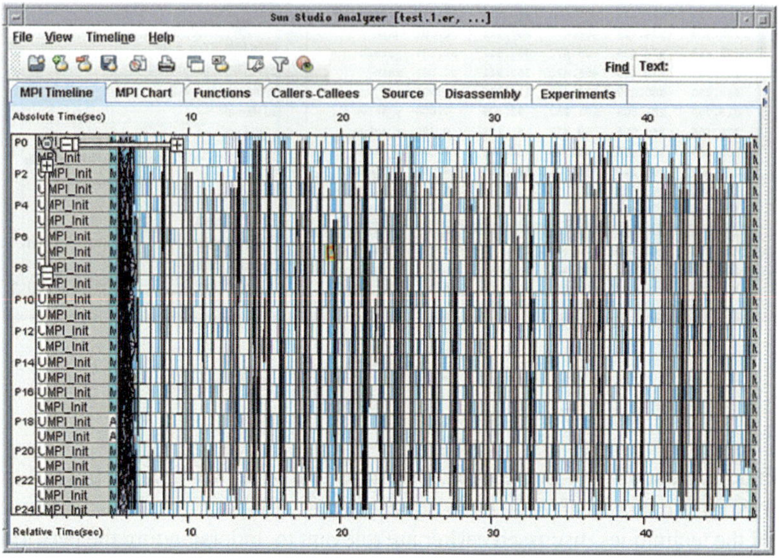

Fig. 6.13: MPI Timeline, Full Scale

The Timeline shows the MPI processes vertically, with time displayed horizontally. For each process, blocks indicating MPI calls and application code are shown. Lines indicating messages are drawn between the sending call and the receiving call for each message. In typical applications, the message volume is quite high, which can lead to a picture that is obscured by the message lines. The user can adjust the display to set a percentage of messages to be shown. Priority of display is given to the most costly messages, that is, the messages that represent the largest amount of time spent in sending and receiving them.

The user can zoom in to help recognize the pattern of execution. The same experiment shown above, when zoomed in, shows a clear pattern (Figure 6.14). In this case, three iterations are shown.

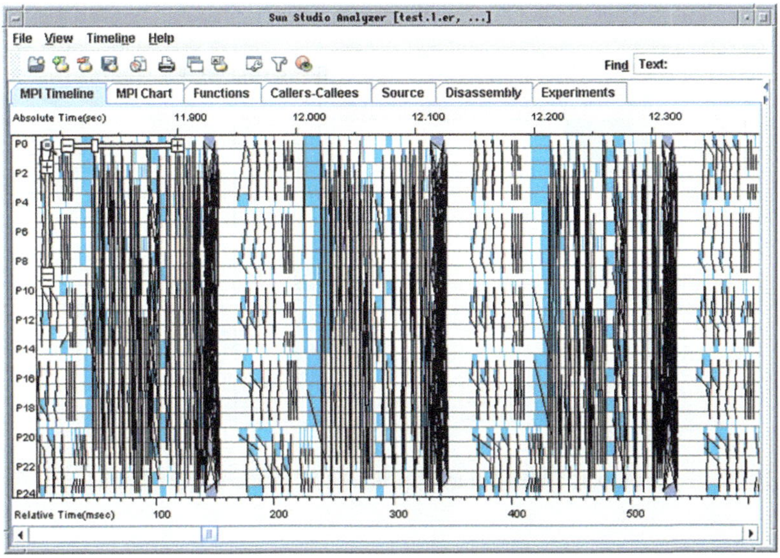

Fig. 6.14: MPI Timeline, zoomed in

The user can zoom in further, and will then see the names of the MPI functions inside their blocks. At any point, the user can select specific MPI events to determine the callstack of the process, and the duration of the call. The user can also select a message to see the message size and the sending and receiving processes and their callstacks.

A filter can be set based on any zoomed-in view of the data, allowing the user to isolate patterns of communication. Typically, the user will set a filter from the MPI Timeline to focus analysis on the steady-state heart of the computations.

6.5.2.2 Using MPI Charts to Understand Where Time Was Spent

The Analyzer's initial MPI Chart shows in which MPI function the time is spent. Figure 6.15 shows the time spent in each of the MPI calls, and in the Application (which is the time spent between MPI calls).

In this example, we can see that approximately 75% of the total time is spent in the application's computation. The remaining 25% of time is spent in MPI calls, with almost all of it spent in the function **MPI_Wait**.

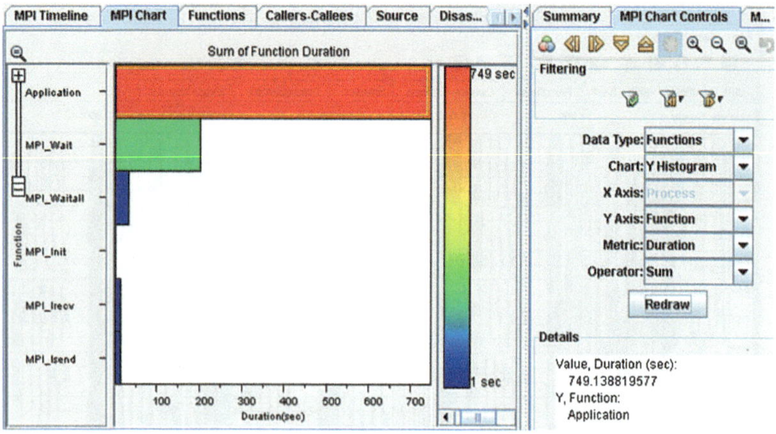

Fig. 6.15: MPI Chart: Application Time vs. MPI Time

6.5.2.3 Using MPI Charts to Understand Message Traffic

The MPI Charts can be used to understand the patterns of communication between processes. In Figure 6.16, we show a 2-dimensional plot, showing data volume in bytes as a color in a grid of sending and receiving processes.

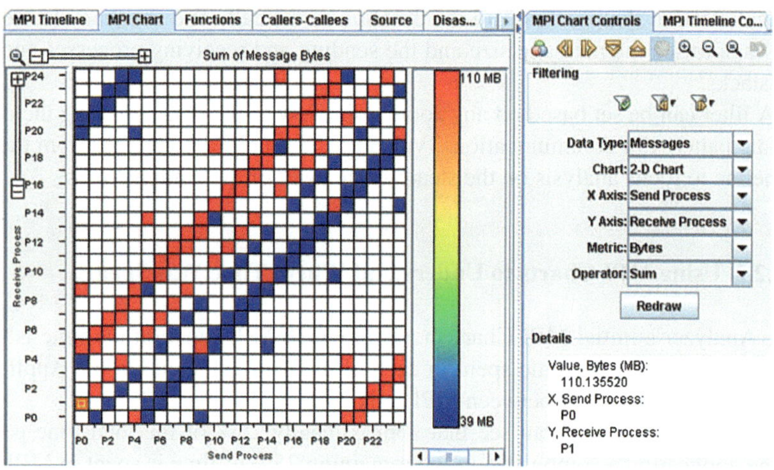

Fig. 6.16: MPI Chart: Bytes-Transmitted among Processes

This chart shows how much data is being passed between the processes. Charts can also be used to explore other aspects of message traffic, including delivery times, and send and receive functions.

6.5.2.4 Using MPI Charts to Understand Work Distribution

The previous techniques have been directed towards understanding the average behavior of the application. They do not indicate if, for example, some processes are running slower than others, or if the behavior is consistent over time. The MPI Charts provide a powerful way to explore these types of issues.

To investigate work distribution, the user can first set a filter to isolate the time in Application, the pseudo-function that represents work done between MPI calls. Then the user can display the amount of time spent in the Application state for each process, as shown in Figure 6.17.

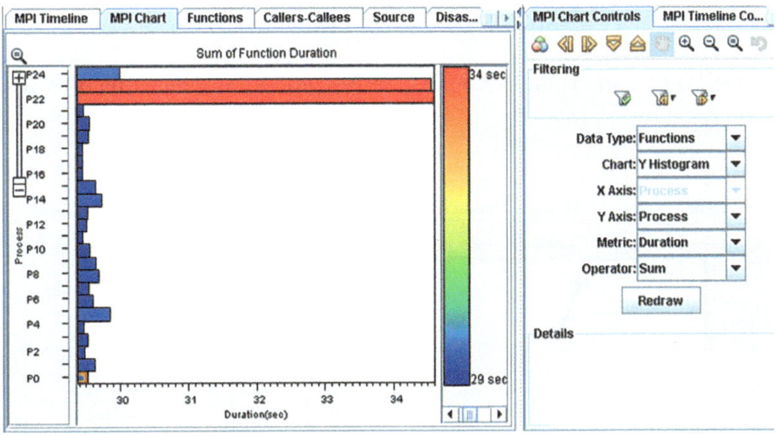

Fig. 6.17: MPI Chart: Time in Application vs. Process

In this example. processes 22 and 23 spend more time in computation than the other processes. Fixing this imbalance may improve the overall performance of the application.

With the filter still set, the user can look at the work distribution over time. Figure 6.18 is a 2-dimensional chart showing process number vertically, wall-clock entry time horizontally, and coloring to represent the relative amount of Application work being done.

The excess time spent in processes 22 and 23, which was visible in Figure 6.17, is now seen to be consistent over the whole run. At the wall-clock times of approximately 19 and 40 seconds into the run, there appears to be hiccups where all the processes are getting less work done.

To investigate time-based anomalies like those shown in Figure 6.18, the user can look at the distribution of Application work periods as a function of wall-clock time, as shown in Figure 6.19.

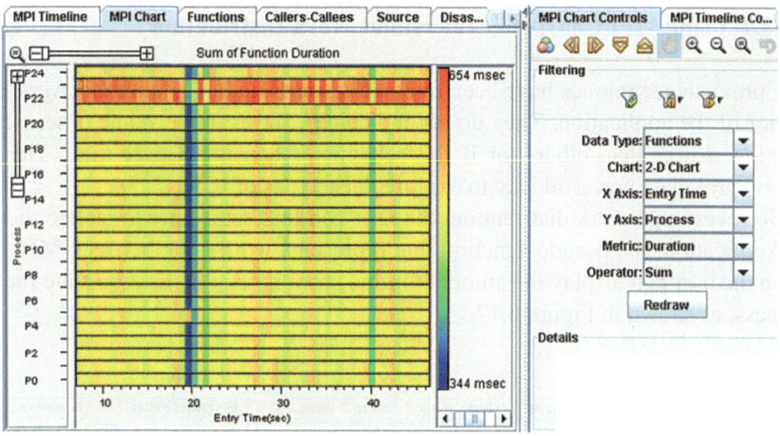

Fig. 6.18: MPI Chart: Time in Application Per-process as a Function of Wall-clock Time

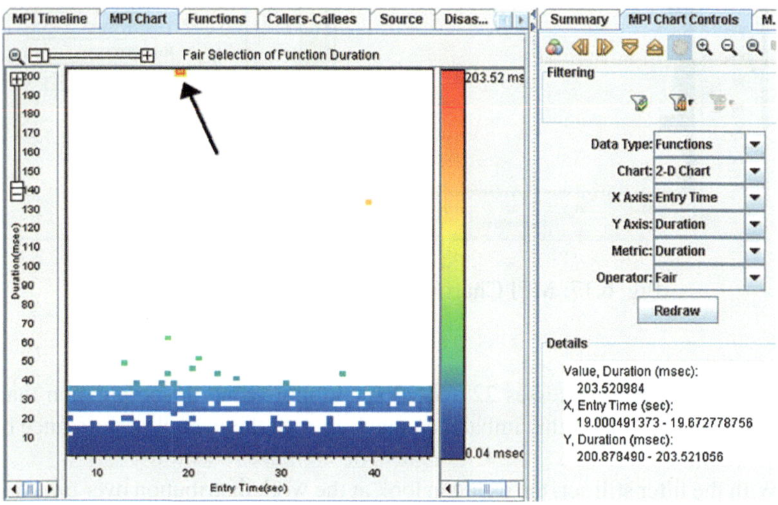

Fig. 6.19: MPI Chart: Time in Application vs. Wall-clock Time

Most work periods (time in Application) are less than 40 milliseconds in duration, but at ~19 seconds into the run, there is a data point showing a work period of 203 milliseconds. There is a second outlier representing the stutter at approximately 40 seconds into the run.

6.5.2.5 Using Filters to Isolate Behaviors of Interest

The MPI filters can be used to pick out behaviors of interest and determine which events are responsible. For example, to focus on that anomalous data point in Figure 6.19, the user can zoom in and apply a filter to isolate the data of interest. Then, the user can switch to the Timeline, zoom in on that event, and remove the filter to show the context of the other processes around that event. Figure 6.20 shows the anomalous event and context on the Timeline.

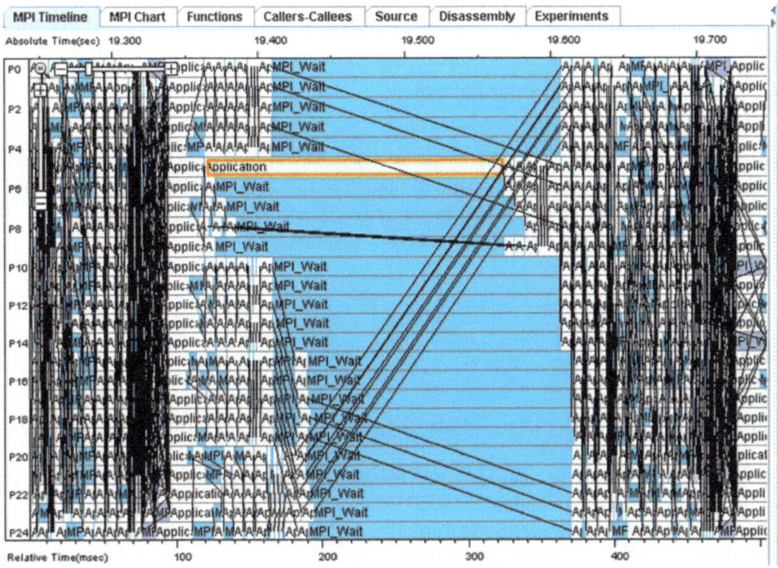

Fig. 6.20: MPI Timeline: Outlier Event Shown

In Figure 6.20, we can see that the long-duration Application event in process 5 impacted all the other processes: while process 5 is computing, all the other processes are waiting. Further drilling down using the Timeline would allow the user to see the source contexts of the anomaly and the surrounding events.

6.6 Conclusions

We have described the Sun Studio Performance Tools and the user model they support. We then discussed single-threaded applications, and the importance of both algorithmic efficiency and memory subsystem behavior to the overall performance of the application. We described the techniques in the performance tools to measure both of these.

We then explored the issues introduced by multi-threading, and gave examples of locking issues, and memory- and cache-contention issues among threads.

We described support in the tools for the OpenMP programming model, and the performance issues concerning OpenMP, including detection of too-little-parallelism, excess-overhead, lock-contention, and load-balance. In each case we showed how the tools can highlight the problems.

Finally, we explored the MPI programming model and the ways in which the tools can measure MPI performance. We described some of the typical characteristics of MPI jobs, and showed how the patterns of communication and computation can be explored. We then showed how the tools can be used to isolate behaviors of interest and to understand their causes.

Acknowledgements We would like to thank Oleg Mazurov, Eugene Loh, Brad Lewis, Nik Molchanov, and Vladimir Mezentsev who did much of the work developing the tools. We would also like to thank some of our users, in particular, Darryl Gove, Ruud van der Pas, Karsten Guthridge, Miriam Blatt, and others too numerous to name.

References

1. Sun Studio Downloads, `http://developers.sun.com/sunstudio/downloads/index.jsp`.
2. S.L.Graham, P.B. Kessler, and M.K.McKusick, An Execution Profiler for Modular Programs, Software Practice and Experience, 13, 671-685, August, 1983.
3. Marco Zagha, Brond Larson, Steve Turner, and Marty Itzkowitz, Performance Analysis using the MIPS R10000 Performance Counters, Proceedings of SuperComputing '96, Pittsburgh, PA, November, 1996.
4. Mikael Pettersson, Linux Performance-Monitoring Counters Driver, `http://user.it.uu.se/~mikpe/linux/perfctr/` Computing Science Division, Uppsala University, Sweden.
5. Stéphane Eranian, "Perfmon2: a standard performance monitoring interface for Linux", `http://perfmon2.sourceforge.net/perfmon2-20080124.pdf`, January 2008.
6. Ingo Molnar, Thomas Gleixner, "[ANNOUNCEMENT] Performance Counters for Linux", `http://lkml.org/lkml/2008/12/4/401`, December 2008.
7. Marty Itzkowitz, Brian J. N. Wiley, Christopher Aoki, and Nicolai Kosche, Memory Profiling using Hardware Counter, Proceedings of SuperComputing '03, Phoenix, AZ, November, 2003.
8. Marty Itzkowitz, Memory Subsystem Profiling with the Sun Studio Performance Analyzer, `http://cscads.rice.edu/workshops/summer09/slides/performance-tools/DProfile.cscads.pdf`.
9. Yuan Lin and Oleg Mazurov. Providing Observability for OpenMP 3.0 Applications, Proceedings of the 5th International Workshop on OpenMP. Dresden (2009).

10. The VampirTrace Project, `http://www.tu-dresden.de/zih/vampirtrace`, Technische Universität Dresden, Center for Information Services and High Performance Computing (ZIH), Dresden, Germany.

Sun, Sun Microsystems, the Sun logo, Solaris and Sun HPC ClusterTools are trademarks or registered trademarks of Sun Microsystems, Inc. or its subsidiaries in the United States and other countries. All SPARC trademarks are used under license and are trademarks or registered marks of SPARC International, Inc. in the U.S. and other countries. Products bearing SPARC trademarks are based upon an architecture developed by Sun Microsystems, Inc.

Chapter 7
Performance Tuning of x86 OpenMP Codes with MAQAO

Denis Barthou, Andres Charif Rubial, William Jalby, Souad Koliai, and Cédric Valensi

Abstract Failing to find the best optimization sequence for a given application code can lead to compiler generated codes with poor performances or inappropriate code. It is necessary to analyze performances from the assembly generated code to improve over the compilation process. This paper presents a tool for the performance analysis of multithreaded codes (OpenMP programs support at the moment). MAQAO relies on static performance evaluation to identify compiler optimizations and assess performance of loops. It exploits static binary rewriting for reading and instrumenting object files or executables. Static binary instrumentation allows the insertion of probes at instruction level. Memory accesses can be captured to help tune the code, but such traces require to be compressed. MAQAO can analyze the results and provide hints for tuning the code. We show on some examples how this can help users improve their OpenMP applications.

7.1 Introduction

Modern processors rely on many complex hardware mechanisms in order to reach high levels of performance. In particular, the use of all levels of parallelism and the appropriate use of the memory hierarchy to hide large memory latencies are both required to obtain the full computing capacity of processors. This road to high performance is paved with many complex compiler optimizations, using, according to the code, prefetching mechanism, vectorization, loop transformations for better cache usage or data layout restructuring. While many optimizing compilers are able to perform all these transformations, they have a poor knowledge of the application

Denis Barthou
University of Bordeaux, LaBRI/INRIA, France

Andres Charif Rubial, William Jalby, Souad Koliai, Cédric Valensi
University of Versailles Saint-Quentin, LRC ITACA, France

M.S. Müller et al. (eds.), *Tools for High Performance Computing 2009*,
DOI 10.1007/978-3-642-11261-4_7, © Springer-Verlag Berlin Heidelberg 2010

context and must be conservative in their transformations. Failing to find the best optimization sequence for a given application code, this leads to compiler generated codes with poor performance, or with inappropriate code.

The performance tuning process therefore implies to guide the compiler, through pragmas, compilation flags, or source to source restructuring, to the generation of better code. Many approaches to performance tuning have been proposed, getting feedback from the application either by collecting execution traces through instrumentation (with Dyninst [4] or Pin [19] for single processors, with Scalasca [20] for multi-node systems) or hardware counters values (such as Intel Vtune or PTU for instance). Hardware counter-based techniques show how the architecture behaves with the considered code and input set. However, it is difficult to make the connection between hardware event counts and source code, since both source code and compiler optimizations have an impact on the resulting hardware events. Moreover, there is no direct link between hardware counters and the quality of the compiler generated code. To have feedback from the compilation process, it is necessary to analyze performance from the assembly generated code.

In this paper, we describe how our MAQAO [6] tool (Modular Assembly Quality Analyzer and Optimizer) handles performance analysis and memory tracing for OpenMP programs. Although in this paper, our target architecture is Core2, the tool can be easily retargeted to other x86 architectures essentially by changing the performance models used. Targeting other architectures requires more work (dealing with different instruction sets) but the main principles can be adapted fairly easily: an earlier version of MAQAO was targeting IA64 architectures which are very different from X86. This tool combines static analysis of compiler-generated assembly code with the analysis of execution traces and binary instrumentation. Static performance evaluation provides hints on how to improve the compilation process, and assess the amount of performance that could be obtained through optimization. This estimation is performed on the sequential codes executed by threads. Improving unicore performance (both in sequential and parallel part of the codes) contributes to improving global performance and efficiency of the code. Dynamic, thread-wise traces, in particular compact memory traces, show how to improve interactions between threads, and detect false sharing situations, for instance. We show in particular how static performance evaluation is achieved on Core 2 architecture and how compact memory traces can be used to help tune OpenMP code performance.

7.2 Static Performance Evaluation

MAQAO relies on static performance evaluation to identify compiler optimizations (or lack of), patterns of codes that are not efficient, and assess performance of loops. The performance model and its use for x86 architecture is described in this section. We first recall how MAQAO analyzes and restructures codes.

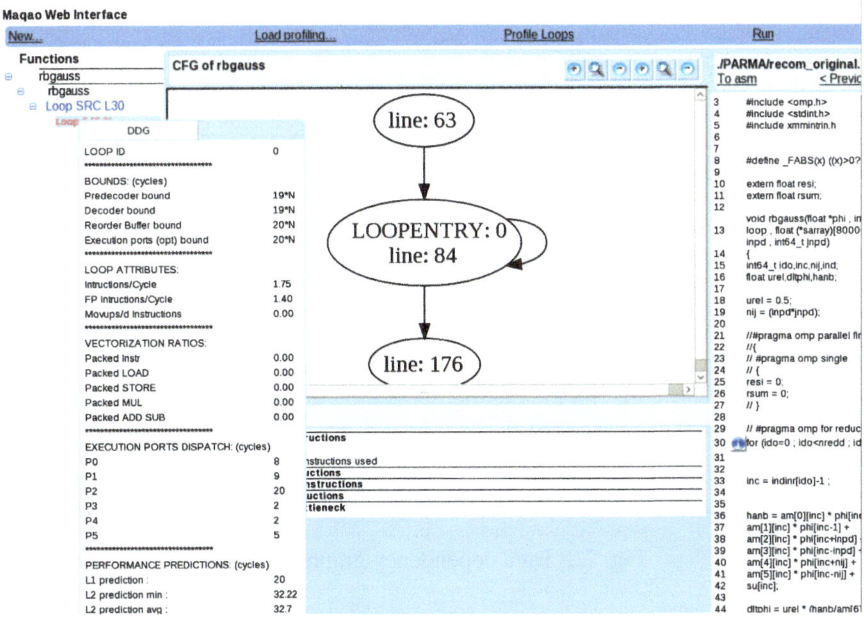

Fig. 7.1: The MAQAO user interface

7.2.1 Code Restructuring

MAQAO exploits static binary rewriting for reading and instrumenting object files or executables. Static binary rewriting refers to the post-link time manipulation of binary executables. This approach has the advantage, compared to approaches requiring compiler interaction (analysis of assembly code) or inclusion of libraries (for heap monitoring for instance), to obviate the need of recompiling or relinking. The API for reading and manipulating static binary files is defined by MADRAS [18], a generic disassembler and instrumenter generator. MADRAS takes a grammar associating binary expressions to assembly instructions, similarly to `yacc` grammars, and generates a corresponding disassembler, using a linear-sweep method (similar to `objdump`). This disassembler for x86 is then used by MAQAO.

The disassembled binary code is restructured: call graphs and control flow graphs, loops and dependence graphs on registers are built (Fig. 7.2). The call graph construction uses labels found in the binary, if any. Both call and control flow graphs are limited in the presence of indirect jumps and self-rewriting codes. So far, there is no (partial) interpretation of the code in order to resolve indirect jumps and self-rewriting of codes. While the first limitation may prevent MAQAO from finding correct control flow, the later may lead to incorrect disassembling. Natural loops are built using a fast algorithm [9].

There is a direct link between each assembly statement and a source code statement provided the debugging information is present (usually given when compiling

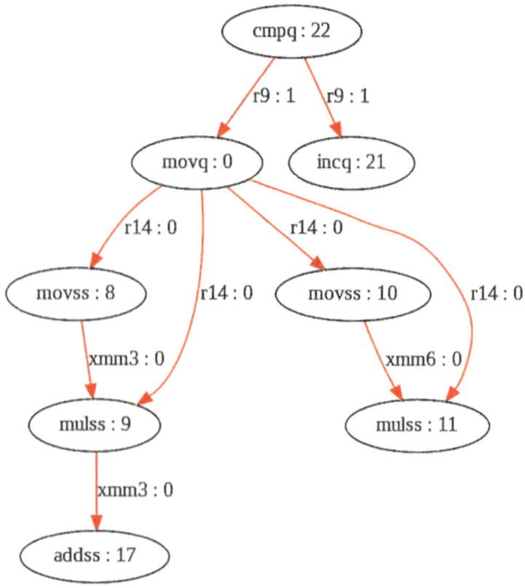

Fig. 7.2: Data dependency graph of a loop

with -g flag). This link allows the detection of some compiler optimizations, such as multiple versioning, inlining and unrolling to some extent. Innermost assembly loops are grouped by source line so that users can visualize the generated assembly loops for a given source loop (Fig. 7.3).

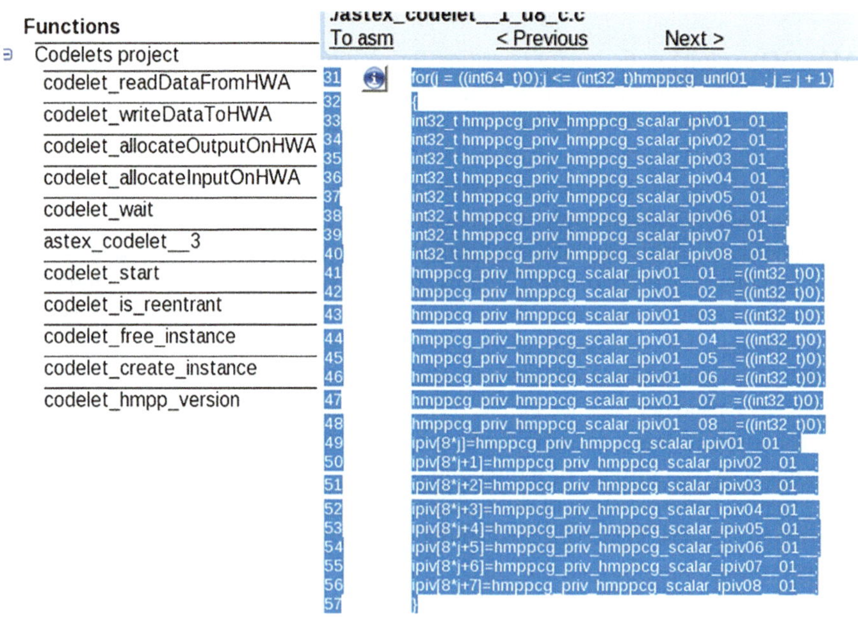

Fig. 7.3: Project - Files - Functions - Loops Hierarchy and corresponding source

7.2.2 Performance Model

The performance model of MAQAO computes performance estimates based on the assembly code. It evaluates the cycles required for executing innermost loops. The reason for considering only the innermost loops is that they usually constitute the most time consuming part of the code. The x86 architecture model we consider takes into account the front-end pipeline (decoding, permanent register file allocation, special microcoded instructions), the different ports for the execution units, and the latencies of instructions. For memory instructions, several latencies are considered, according to the location of the data in memory hierarchy. For other instructions, latencies are tabulated, either coming from microbenchmarks or from Agner documentation [8]. Note that the evaluation only provides an optimistic bound, meaning that the real code may execute in more cycles due to some extra latency not taken into account by our model.

Among different metrics that MAQAO can produce, we focus on the following five key metrics:

1. **Vectorization Report Analysis:** This report, shown in Fig. 7.5b, provides us with individual (load, store, add, multiply) reports on vector instruction usage: for example a vector ratio of 1 for multiply operations means that all of multiply operations have been vectorized by the compiler. This ratio is computed taking into account only floating point operations and full length packed vector operations. These metrics are essential to evaluate the quality of the vectorizing capabilities of the compiler and possibly to palliate some of its deficiencies by inserting appropriate pragmas.
2. **Execution port usage:** For each execution port (Fig. 7.4), MAQAO computes an estimation of the number of cycles spent on each port. Our performance estimates takes into account the special case of instructions which are split into

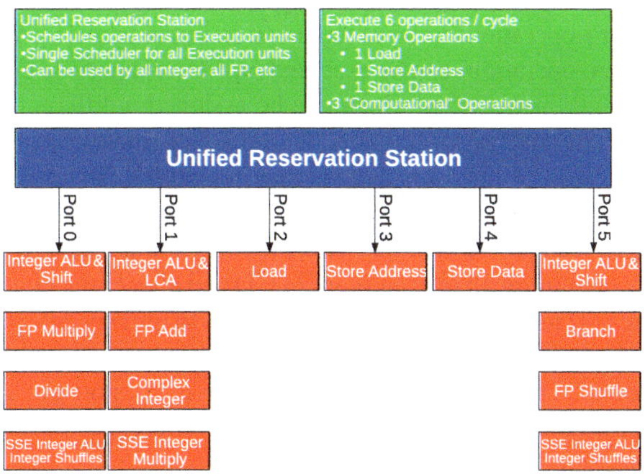

Fig. 7.4: Core2 execution unit overview

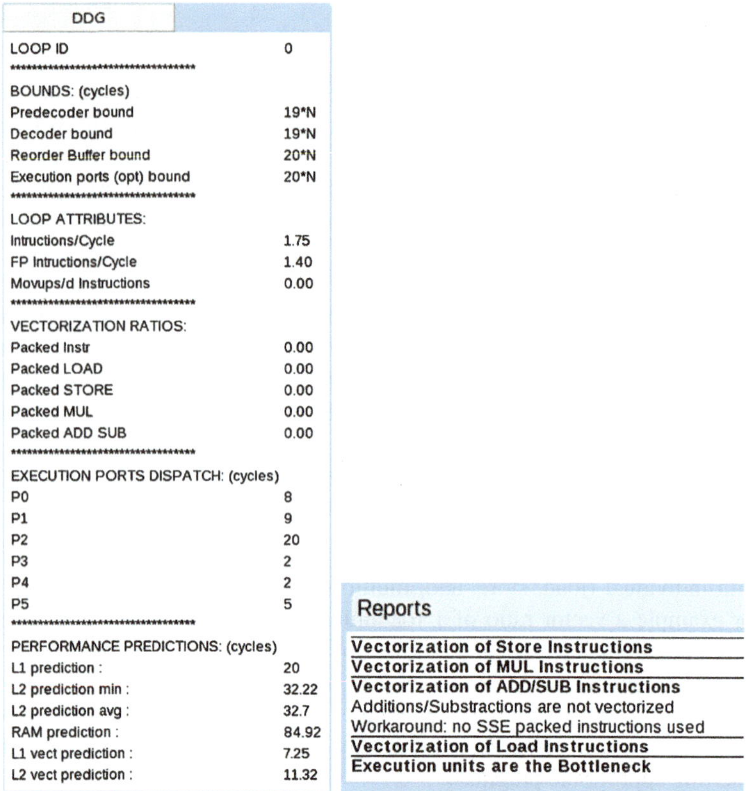

(a) MAQAO statistics (b) MAQAO reports

Fig. 7.5: MAQAO interface details.

different micro-operations to be executed on multiple ports [8]. When an instruction (or a micro operation) can be executed on different ports (a common example is simple integer instructions which can be assigned indifferently to P0, P1 and P5), the less saturated port is chosen. Figure 7.5a shows the report presented by MAQAO Since all of the ports can operate in parallel, this metric is essential to measure the amount of parallelism exploitable between the key functional units: add, multiply, load and store units. This provides a first estimate of a best performance case (assuming all operands are in L1) and also of the potential imbalance between the port usage. For example, this allows to quickly detect whether a code is memory bound and to get a first quantitative estimate of how much a code is memory bound. The number of cycles spent on every port gives us an accurate ranking on the potential bottlenecks of the code difference in cycles between first order and second order bottlenecks).

3. **Performance estimation in L1:** Taking into account all of the limitations of the pipeline front end and of the pipeline back end, MAQAO provides us with

an estimate of the cycles necessary to execute one loop iteration assuming all operands are in L1. The limitations that we are taking into account are: instruction predecoding, instruction decoding, permanent register file allocation, special microcoded instructions. As mentioned earlier, in most cases this bound is only useful as a lower bound.

4. **Performance estimations in L2/RAM:** Relying on memory access patterns detected at the assembly level and micro benchmarking results on the same memory patterns, MAQAO computes an estimate for the execution time of a loop iteration, assuming all operands are in a given level of the memory hierarchy (L2 or RAM) and are accessed with stride 1. The memory patterns used for the pattern matching have previously been determined by systematic hierarchical microbenchmarking: first simple "Load X" (resp. "Store Y") kernels (performing a single read stream through an array X, resp. a simple writing stream through an array Y) are measured under various conditions (unrolling, instruction used, etc ..). Then more complex patterns "Load X Store Y", "Load X Load Y", "Load X Load Y Store Z", etc ... are measured to quantify the interaction between Load streams and Stores streams. We experimentally observed that beyond 4 array streams, most of the performance measured could be deduced from simpler patterns. Therefore this simple set of patterns is used for our performance prediction [10]. The L2 estimate constitutes a reasonable performance objective while the RAM estimate is a stride 1 worst case. The drawback of both of these estimates is that they ignore the stride problem (which in RAM will be essential) and, second, that they do not take into account the mixture of hits and misses which is typical for real applications. However, it should be noted that micro benchmarking already accounts for some typical mixture of hits/miss resulting from spatial locality usage. For stride 1 memory access, micro benchmarking does not distinguish between primary misses (occurring for the first word access to a cache line) and secondary misses/hits (occurring when subsequent words in the cache line are requested), it provides an estimate of the average time for accessing a memory location in a stride 1 access mode (array stored in contiguous memory). The stride problem can be easily corrected when the memory tracing analysis is performed, because for each load/store, the striding pattern will be then determined. Then a revised more accurate L2/RAM estimate can be generated. Again incorporating this extra information enables MAQAO to produce better performance estimates.
5. **Performance projections for full vectorization:** In cases where the code is partially or not vectorized, MAQAO computes performance estimations assuming a full vectorization. This is performed by replacing the scalar operations by their vector counterparts and updating the timing estimate due to the use of these instructions. This is particularly useful to guide the optimization process and to avoid useless efforts: for example, indirect access to arrays cannot be vectorized due to the lack of vector scatter/gather instructions in the current SSE instruction sets. However, in most loops, these indirect accesses are followed by floating point operations (adds or multiplies) which could be vectorized. The MAQAO performance projection gives us quickly an estimate of whether trying to vectorize these operations will pay off or not.

7.2.3 Applying MAQAO to Real-World Applications

To illustrate the interest of these metrics, we performed a static analysis using MAQAO on two high performance codes from the ParMA project [16]: `RECOM-AIOLOS` from RECOM, and `ITRLSOL` from Dassault-Aviation. Two code fragments are shown in Fig. 7.6. The Intel C and Fortran Compilers (ifort and icc v11.0) are used to generate the assembly codes analysed by MAQAO. They are also used to generate OpenMP parallel regions when appropriate and also all of the performance measurements have been carried out using these compilers.

```
DO IDO=1,NREDD
  INC    =    INDINR(IDO)
  HANB   =    AM(INC,1)*PHI(INC+1)   &
  + AM(INC,2)*PHI(INC-1)   &
  + AM(INC,3)*PHI(INC+INPD)   &
  + AM(INC,4)*PHI(INC-INPD)   &
  + AM(INC,5)*PHI(INC+NIJ)   &
  + AM(INC,6)*PHI(INC-NIJ)  &
  + SU(INC)
  DLTPHI = HANB/AM(INC,7)-PHI(INC)
  PHI(INC) = PHI(INC) + DLTPHI
  RESI = RESI + ABS(DLTPHI)
  RSUM = RSUM + ABS(PHI(INC))
```

(a) `RECOM-AIOLOS` analyzed code fragment

```
DO cb=1,ncbt
  igp = isg isg = icolb(icb+1) igt = isg - igp
c$OMP  PARALLEL DO DEFAULT(NONE)
c$OMP  SHARED(igt,igp,nnbar,vecy,vecx,ompu,ompl)
c$OMP  PRIVATE(ig,e,i,j,k,l)
  DO ig=1,igt
    e = ig + igp
    i = nnbar(e,1)
    j = nnbar(e,2)
cDEC$ IVDEP
   DO k=1,ndof
cDEC$ IVDEP
      DO l=1,ndof
        vecy(i,k) = vecy(i,k) + ompu(e,k,l)*vecx(j,l)
        vecy(j,k) = vecy(j,k) + ompl(e,k,l)*vecx(i,l)
```

(b) `ITRLSOL` analyzed code fragment

Fig. 7.6: Two examples of codes. The `IVDEP` pragma tells the compiler to vectorize the loops.

The different execution ports P0 to P5 in the Core2 architecture correspond to (Fig. 7.4):

- P0-P1-P5: computation units port
- P2: memory read port
- P3-P4: memory write ports

Depending on the number of cycles spent in each port, this information allows to detect if the code is memory bound (P2, P3-P4) or compute bound (P0-P1-P5).

The 3D-combustion modeling software `RECOM-AIOLOS` is a tailored application for the mathematical modelling of industrial firing systems ranging from several hundred kW to more than 1000 MW. In-depth validation using measurements from industrial power plants, the extension of chemical reaction models and the rapid development of computer technology have made RECOM-AIOLOS a well proven and reliable tool for the prediction equations on a 10-15 million cells finite volume grid, leading to high computational demands. Originally being designed for high-performance computing on parallel vector-computers and massively parallel systems, the software has been ported to low-cost multi-core systems to expand the hardware base [17].

The most time consuming subroutine in `RECOM-AIOLOS` is `RBgauss`, which implements a red-black iterative solver. The choice of the red-black algorithm allows for easy parallelization with, for example, OpenMP. The `RBgauss` subroutine contains two loops (denoted *Red* and *Black* loop) with a communication between them using MPI. The static analysis with MAQAO is performed on the *Red* loop as both loops are the same. It gives the following values:

- Vectorization report: all the ratios of vectorization are equal to 0%. The compiler has not vectorized the loop.
- Execution units usage (format is PORT_NUMBER:CYCLES_SPENT): P0:8 / P1:10 / P2:19 / P3:1 / P4:1 / P5:4.
- L1 prediction: 19 cycles.
- L2 prediction: 28.77 cycles.
- RAM prediction: 70.66 cycles.
- Vectorization prediction (assuming data in L1): 7 cycles.

Thanks to the static analysis of MAQAO, we can notice that the code is memory bound on Core 2, since it takes 19 cycles to execute all read instructions. This corresponds to the largest number of cycles on any given port.

The memory traces achieved using MAQAO allowed to detect that there are two arrays (`AM` and `PHI`) in the code which are accessed with a stride 2 with some gaps from time to time.

Moreover, the large number of reads and the stride 2 access imply that the code is very sensitive to cache misses [12].

Since the major bottleneck for this routine is data access from RAM combined with low spatial locality (stride 2 access), various optimizing transformations are performed, but only the following has a significant impact on performance: reshaping array AM for getting rid of the stride 2 access. More precisely, the array `AM` is split into two distinct arrays still with indirect access but stride 1. This is equivalent to reshaping an array of complex numbers by splitting it into arrays, one containing the real part, the other one containing the imaginary part.

Thanks to this optimization, the cache misses are almost half what they used to be (Fig. 7.7b). Single core performance has been improved by speedups between

1.2 and 1.3 (Fig. 7.7a) thanks to this code transformation. Multicore performance has been improved by speedups between 1.3 and 1.4 (Fig. 7.8).

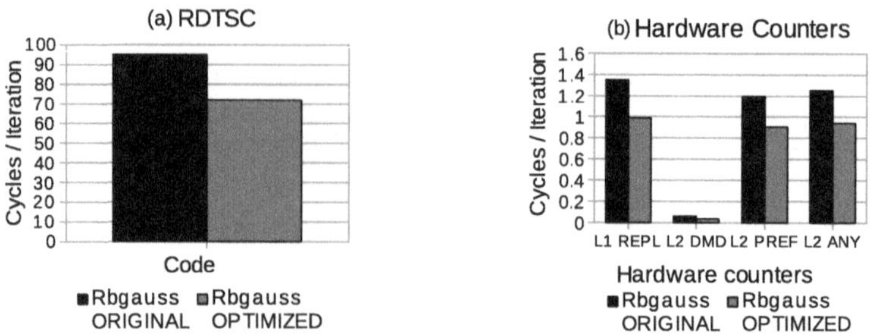

Fig. 7.7: `RBgauss` code optimization on unicore.

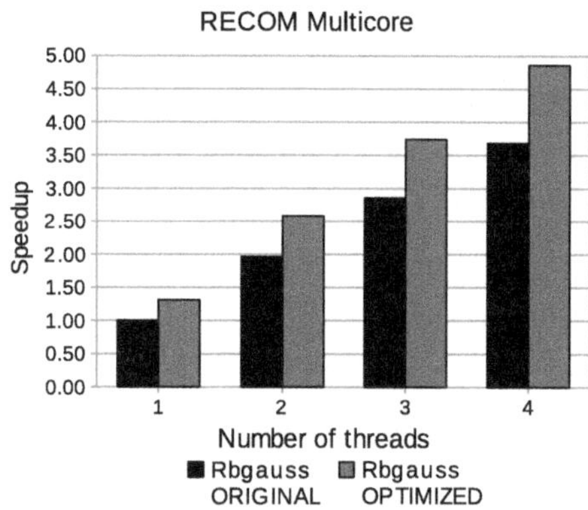

Fig. 7.8: `RBgauss` speedups on multicore.

The `ITRLSOL` (ITeRative Linear SOLver) application provided by Dassault-Aviation is the linear solver kernel of AeTHER, a larger Computational Fluid Dynamics (CFD) simulation code for the solution of Navier-Stokes equations, discretized on unstructured meshes. The most time-consuming subroutine in `ITRLSOL` is `EUFLUXm`, which implements a sparse matrix-vector product. The `EUFLUXm` subroutine contains two groups of quadruply nested loops (2 identical quadruply nested loops in each group). For the considered 4-level loop nest in this code, the report provides the following information:

- Vectorization report: all the ratios of vetorization are equal to 0%. The compiler has not vectorized any loop, despite the presence of pragmas.

- Execution units usage (format is PORT_NUMBER:CYCLES_SPENT): P0:3 / P1:3 / P2:6 / P3:2 / P4:2 / P5:3
- L1 prediction: 6 cycles.
- L2 prediction: 9.08 cycles.
- RAM prediction: 37.04 cycles.
- Vectorization prediction (assuming data in L1): 3 cycles.

The static analysis with MAQAO shows that the code is dominated by memory accesses. The memory traces achieved with MAQAO allow us to detect that the inner most loops are accessing the arrays in the wrong dimension which leads to a poor spatial locality [12].

To improve the spatial locality, a transformation is done by interchanging the second loop on *ig* and the two innermost loops (the *ig* loop becomes the innermost loop). All of the arrays are now accessed column-wise. This optimization improves sequential performance by speedup of 2.5 (Fig. 7.9a, Fig. 7.9b).

In a multicore environment the same optimization is applied. It gives a speedup of up to 2.5 (Fig. 7.10).

Thanks to the information collected from the static analysis with MAQAO, we detect that both applications `RECOM-AIOLOS` and `ITRLSOL` are not vectorized and memory bound. Using this information and applying MAQAO memory traces and PTU [3] (for performance tuning) allows us to find the performance bottleneck (stride 2 access for `RECOM-AIOLOS` and poor spatial locality for `ITRLSOL`) in these codes.

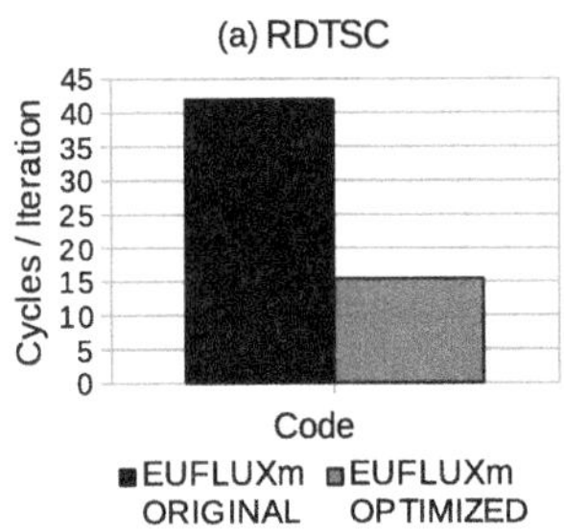

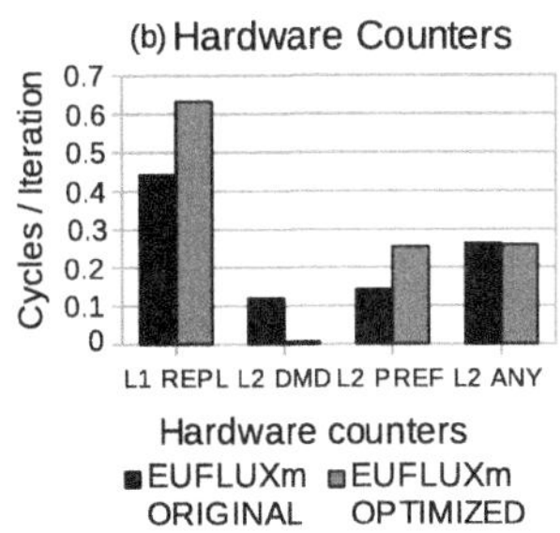

Fig. 7.9: `EUFLUXm` code optimization on unicore.

7.3 Memory Traces for OpenMP Codes

Memory traces represent information of crucial importance for performance tuning of multithreaded codes. Indeed, traces can help detect important inefficiencies (false sharing) or opportunities for optimizations (setting thread affinity according to reuse among threads). The major issue of memory traces is the amount of data they repre-

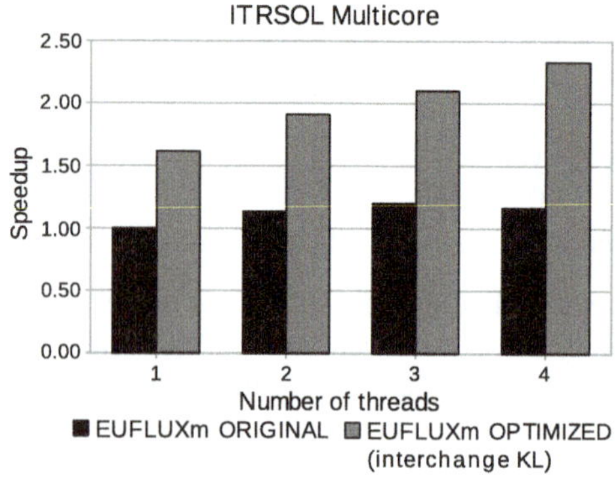

Fig. 7.10: ITRLSOL speedups on multicore.

sent. We first describe how the tracing is achieved in MAQAO, which algorithm we use to compress the traces and how they are used in order to tune performance.

7.3.1 Static Binary Instrumentation

The static binary instrumentation is achieved using MADRAS [18]. It allows instruction level instrumentation, inserting probes either provided by MAQAO (for iteration counts) or user-defined ones in libraries.

Figure 7.11 shows how easy it is to use this API to build an instrumenter module. The two *for* loops walk through all blocks and all instructions of the loop with

```
loopid = 1;
instru = madras.new(binfilename);
for block in loop_blocks(loop_table[pname][loopid]) do
    for ins in instructions(block) do
        if( (ins:is_load() or ins:is_store()) ) then
            instru:fctcall_new("mt_store","libmaqaotrace.so",ins:get_address(),0);
            instru:fctcall_addparam_imm(instru_count);
            if(ins:is_load()) then
                instru:fctcall_addparam_frominsn(ins:get_operand_src_index()-1,
                        ins:get_address());
            else
                instru:fctcall_addparam_frominsn(ins:get_operand_dest_index()-1,
                        ins:get_address());
            end
            instru_count = instru_count + 1;
            instru_tab[ins:get_address()] = instru_count;
        end
    end
end
instru:modifs_commit(binfilename.."_".."instrumented");
instru:terminate();
```

Fig. 7.11: MADRAS API available through MAQAO (LUA scripting interface)

id 1. For each load and store instruction, the `mt_store` function is called from the `libmaqaotrace` library which contains the implementation of the trace compression algorithm mentioned earlier taking into account multithreading. This function builds a compact trace of memory accesses. MADRAS performs instrumentation statically, through binary rewriting, allowing the instrumented program to be run without additional overhead.

7.3.2 Memory Traces

Memory accesses can dramatically slow down the execution time of a program, particularly when it is memory bound. Capturing the memory behavior of a program can help tune the code, using prefetching or transforming the code for a better reuse of data. However, tracing memory accesses (load, store, prefetch) by simply dumping all address streams would lead to many Terabytes of data on real applications. The memory space for these traces is a major concern in every trace-profiling application. We first detail the compression algorithm used in MAQAO and then describe how this method has been adapted to MAQAO for tracing multithreaded codes (OpenMP programs support at the moment).

Compression Algorithm

The compression is ensured by an on-the-fly incremental algorithm called loop nested recognition and developed by Ketterlin and Clauss [11]. We recall in this section the main steps of this method.

Their technique represents memory address streams as union of Z-polytopes which are represented by (nested) loops. The idea of using loops to characterize an accessed region has first been introduced by Elnozahy [7]. Simpler representations have been proposed using triplets (starting addresses, stride, number of references) and their extension to multidimensional triplets [13]. This is a natural approach since the majority of time execution of a program is spent in loops, and memory accesses are regular. Figure 7.12 shows the parallel between a typical example of program loop and its representation.

C code	**Corresponding Trace**
`for (j=0; j<NCB; j++)....`	for i0 = 0 to 7
`    for(i=0; i<NRA; i++)`	for i1 = 0 to 63
`.       for (k=0; k<NCA; k++)`	for i2 = 0 to 63
`.               c[i*NRA+j] += ...;`	val 0x62cd5a0 + 64*i0 + 512*i1

Fig. 7.12: Source code loop and its corresponding nested loop representation

The algorithm takes into account two types of access patterns:

- regular patterns obtained by regular or irregular accesses
- irregular patterns due to random accesses. There is no easy way to deal with this kind of pattern. Existing approaches fall back on lossy algorithms.

Each memory stream is assigned an internal stack that stores either regular and irregular patterns. Regular patterns are stored in the loop format described above. Irregular patterns, which correspond to a sequence of numbers without any affinity, are kept as it is. The stack size management is controlled by three factors:

- the maximum stack size (length)
- the maximum number of terms within the loop body representation (breadth)
- the number of elements to throw when the size limit is reached

The algorithm is lossless as long as the stack is large enough to store all memory streams. On some huge programs it may be necessary to voluntarily limit the stack in order to prevent consuming all the available memory. In this case the algorithm is lossy.

Multithread and Performance Issues

We have adapted the previous method to a multicore execution context and extended it by taking into account static analysis information. Adaptation to multicore execution boils down to reimplementing the original method as a thread-safe method. Traces are saved for each memory access, for each thread independently.

Instrumenting a code for memory traces usually generates a large overhead, and most methods (such as Metric[13]) use sampling in order to reduce this weight on the execution time. Another approach is to use static analysis in order to infer fragments of traces from the assembly code. Indeed, tracking down the induction variable of inner loops makes it possible to capture the stride of memory streams. The value may appear in the code as a numerical constant or as a parametric constant (invariant in the loop). But in both cases we only need this value and the iteration count to extract the loop representation (usually found in some register). Thus this saves a large part of the overhead due to instrumentation.

Our design still suffers from a lack of information about temporal locality. This could be alleviated by using a simplified cache simulator. Its integration in MAQAO is left for future work.

7.3.3 Using Traces for OpenMP Performance Issues

Once trace collection is done, the results are analyzed (manually at the moment) and some hints are provided to the user to help tune the code. We provide thereafter a number of scenarii reflecting performance issues that can be detected using this trace framework.

- Potential bank/load store queue conflicts: this type of conflicts can be easily detected by comparing addresses accessed by "neighbor" instructions. On Xeon architecture a store on address A followed by a load on address B can generate pipeline stalls if address A and B have the same low order 12 bits (same offset within a page). The performance impact will depend upon the execution distance (how many cycles apart) between the load and store instructions. Detection of this intra-thread issue consists in finding successive load/store patterns accessing different addresses sharing the same low order 12 bits.
- False Sharing: Two threads share some cache line, while they do not share any data. However, performance is impacted due to cache coherency issues. Patterns that lead to false sharing can be tracked down by comparing addresses read and written by different threads (loads and stores). Coherence issues increase with the number of cores and the memory access is not uniform (case of Intel architectures).
- Prefetch distance: Prefetch distances can be found or guessed based on the accessed regions of memory. The memory region found by our trace mechanism helps the user to determine if prefetch causes potential false sharing issues, depending on the prefetch distance.
- OpenMP work distribution scheme: Based on the memory pattern accesses, we can recover OpenMP work distribution scheme (different static, dynamic and guided modes) and evaluate which mode is the more appropriate for the application;
- Reuse degree between loads: Multiple loads on the same, shared data give us temporal locality hints. The user could, if possible, reorder some statements to take advantage of cached data at some point;
- Strided accesses: Depending on the programming language, data is stored by column or by row in memory. One possible optimization is to assess which configuration is the most efficient. Moreover, structure of arrays or arrays of structures are usual choices that impact performance, in particular due to vectorization. Evaluation from traces of the opportunity to vectorize memory accesses is an important task in the code tuning phase.

To illustrate one of these scenarii, consider the code shown in Fig. 7.12. This code could match for instance to a matrix multiplication code. The i loop can be parallelized with OpenMP, and different load balancing methods can be chosen, among which `STATIC` and `DYNAMIC` methods. Tracing memory writes reveals that with a `STATIC` load balancing method and 8 threads, there is no false sharing occurring. For the `DYNAMIC` method, as shown in Fig. 7.13, some false sharing occurs, resulting in increased memory latency due to cache coherency mechanism (false sharing for write accesses).

```
Thread 1                      Thread 4                      Thread 7
for i0 = 0 to 127             for i0 = 0 to 127             for i0 = 0 to 127
  for i1 = 0 to 127             for i1 = 0 to 127             for i1 = 0 to 127
    val 0x45255b0 + 1024*i0       val 0x45255d8 + 1024*i0       val 0x45255c8 + 1024*i0

Thread 2                      Thread 5                      Thread 8
for i0 = 0 to 127             for i0 = 0 to 127             for i0 = 0 to 127
  for i1 = 0 to 127             for i1 = 0 to 127             for i1 = 0 to 127
    val 0x45255b8 + 1024*i0       val 0x45255d0 + 1024*i0       val 0x45255a0 + 1024*i0

Thread 3                      Thread 6
for i0 = 0 to 127             for i0 = 0 to 127
  for i1 = 0 to 127             for i1 = 0 to 127
    val 0x45255c0 + 1024*i0       val 0x45255a8 + 1024*i0
```

Fig. 7.13: Partial traces corresponding to memory writes in a DGEMM code, where the i loop is parallelized with `DYNAMIC` load balancing strategy. This solution is not efficient due to the false sharing between threads (for instance, threads 1 and 2 access the same cache line).

7.4 Related Work

There are few performance tools dealing with parallel (multithreaded) codes optimization.

Intel VTune [5] relies on the Thread Profiler application to determine the number of cores that are being used, show the distribution of work to threads but does not take into account memory accesses.

Acumem [1] rely on cache related statistics to predict performance bottlenecks. MAQAO uses (memory accesses) tracing rather than sampling in order to provide the user with very accurate results and detect unusual behaviours.

HPCToolkit [2] also works at binary level for language independence, collects and correlates multiple performance metric, computes derived metrics to aid analysis. However it uses profiling rather than adding code instrumentation. MAQAO supports code instrumentation to enable users inserting probes and concentrate on specific parts of an application.

PIN [19] and DynInst [4] are two tools allowing modification of an executable for the purpose of instrumentation. Both perform dynamic instrumentation, operating on the executable while it is loaded in memory and running.

PIN traces an executable during its execution (acting as a "just-in-time" compiler) and monitors various parameters. It allows to transfer control flow to external functions, effectively inserting calls to these functions, and to modify the memory, all of this while the executable is being executed. It is also possible with PIN to insert probes in the executable while it is loaded in memory but not yet running, which actually redirects the execution flow to another function. This mode does not work on multi-threaded applications and does not check the destinations of jump instructions. PIN is also able to perform some static analysis of a file (like identifying functions arguments).

DynInst [4] allows dynamic updating of code, a process labeled runtime injection. It proceeds by directly updating a program in memory to insert jumps pointing to the added sections of code which reside somewhere else in memory. A recent update also allows DynInst to perform some binary rewriting.

Instrumentation by MADRAS is not accomplished at runtime by another thread, as it is performed statically. The instrumented program can then be run without additional overhead but the calls to the instrumented functions. No special environment is required.

The integration of the MADRAS library allowed us to introduce the memory tracing feature.

Valgrind [14] is a dynamic binary instrumentation and analysis framework which uses a simulated CPU to analyse programs (in particular on cache and memory use) and offer instrumentation options. The simulated CPU causes an important slowdown of the analysed program and requires more memory space.

METRIC [13] uses dynamic instrumentation to capture memory accesses and scope changes.

PSnAP [15] also uses dynamic instrumentation to generate memory stream profiles on a per loop basis as MAQAO does.

Ketterlin & Clauss [11] propose a more sophisticated compression technique that we are using in our memory tracing library.

To our best knowledge, there is no existing technique for memory tracing of parallel (multithreaded) codes.

7.5 Conclusions and Future Work

MAQAO is a tool for performance tuning that relies on both static analysis of binaries and on data collected through instrumentation. We have shown in this paper how the performance model for x86 processors is designed inside MAQAO and how memory tracing for OpenMP programs is achieved.

The static analysis is combined with the hint mechanisms of MAQAO, helping the user to locate easily in the application source code the code fragments that exhibit poor performance. Moreover, this analysis provides a rough estimate of the possible performance gains that could be expected by an efficient vectorization. The memory tracing method we propose relies on two mechanisms: a new binary instrumentation framework, MADRAS, where each assembly instruction can be instrumented individually, and a compact memory trace representation [11], extended for multithreaded programs. We have shown, through multiple scenarii, how the multithreaded trace information can be used to detect performance issues specific to multicore machines.

For future work, we plan to improve the trace representation in order to capture partially some scheduling information (associating time stamps with memory addresses). In future versions, trace results will be analysed automatically. MAQAO still needs the assembly code for building its analysis and will rely only on the data extracted from the disassembled binary in the next release (the disassembled binary from MADRAS being used in correlation with it to retrieve the instructions addresses).

References

1. Acumum AB. Acumem SlowSpotter and Acumem ThreadSpotter, 2009. http://www.acumem.com/content/view/133/182/.
2. L. Adhianto, S. Banerjee, M. Fagan, M. Krentel, G. Marin, J. Mellor-Crummey, and N. R. Tallent. HPCToolkit: Tools for performance analysis of optimized parallel programs. Technical Report TR08-06, Rice University, 2008.
3. A. Alexandrov, S. Bratanov, J. Fedorova, D. Levinthal, I. Lopatin, and D. Ryabtsev. Parallelization Made Easier with Intel Performance-Tuning Utility, 2007. http://www.intel.com/technology/itj/2007/v11i4/.
4. B. Buck and J. K. Hollingsworth. An API for Runtime Code Patching. *Intl. Journal of High Performance Computing Applications*, 14:317–329, 2000.
5. Intel Corporation. Intel VTune Performance Analyzer 9.1, 2009. http://software.intel.com/en-us/intel-vtune/.
6. L. Djoudi, D. Barthou, P. Carribault, C. Lemuet, J-T. Acquaviva, and W. Jalby. Exploring Application Performance: a New Tool For a Static/Dynamic Approach. In *Los Alamos Computer Science Institute Symp.*, Santa Fe, NM, October 2005.
7. E. N. Elnozahy. Address trace compression through loop detection and reduction. *SIGMETRICS Perform. Eval. Rev.*, 27(1):214–215, 1999.
8. Agner F. Software optimization resources, 2009. http://www.agner.org/optimize/.
9. L. Georgiadis, R. F. Werneck, R. E. Tarjan, S. Triantafyllis, and D. I. August. *Algorithms - ESA*, 3221:677–688, 2004.
10. W. Jalby, C. Lemuet, and X. Le Pasteur. A New Set of Microbenchmarks to Explore Memory System Performance for Scientific Computing, 2004. International Journal of High Performance Computing Applications.
11. A. Ketterlin and Ph. Clauss. Prediction and Trace Compression of Data Access trough Nested Loop Recognition. In *ACM/IEEE Int. Symp. on Code Optimization and Generation*, 2008.
12. S. Koliai, S. Zuckerman, E. Oseret, M. Ivascot, T. Moseley, D. Quang, and W. Jalby. A Balanced Approach to Application Performance Tuning. In *Proc. of LCPC*, LNCS, Delaware, USA, October 2009. Springer.
13. J. Marathe, F. Mueller, T. Mohan, B. R. de Supinski, S. A. McKee, and A. Yoo. METRIC: Tracking Down Inefficiencies in the Memory Hierarchy via Binary Rewriting. *ACM/IEEE Int. Symp. on Code Optimization and Generation*, 0:289, 2003.
14. N. Nethercote and J. Seward. Valgrind: A Framework for Heavyweight Dynamic Binary Instrumentation. 2007. Proceedings of ACM SIGPLAN 2007 Conference on Programming Language Design and Implementation (PLDI 2007), San Diego, California, USA, June 2007.
15. C. Mills Olschanowsky, M. Tikir, L. Carrington, and A. Snavely. PSnAP: Accurate Synthetic Address Streams Through Memory Profiles. In *Int. Workshop on Languages and Compilers for Parallel Computing*, 2009.
16. ParMA ITEA2 Project: Parallel Programming for Multicore Architectures. http://www.parma-itea2.org/.
17. B. Risio, A. Berreth, S. Zuckerman, S. Koliai, M. Ivascot, W. Jalby, B. Krammer, B. Mohr, and T. William. How to Accelerate an Application: a Practical Case Study in Combustion Modelling. In *Proc. of ParCo*, Lyon, France, 2009.
18. C. Valensi and D. Barthou. MADRAS: Multi-Architecture Disassembler and Reassembler, 2009. http://maqao.prism.uvsq.fr/wiki/wiki/MadrasDownload.

19. S. Wallace and K. Hazelwood. SuperPin: Parallelizing Dynamic Instrumentation for Real-Time Performance. In *ACM/IEEE Int. Symp. on Code Optimization and Generation*, pages 209–217, San Jose, CA, March 2007.
20. F. Wolf, B.J.N. Wylie, E. Ábrahám, D. Becker, W. Frings, K. Fürlinger, M. Geimer, M.-A. Hermanns, B. Mohr, S. Moore, M. Pfeifer, and Z. Szebenyi. Usage of the SCALASCA Toolset for Scalable Performance Analysis of Large-Scale Parallel Applications. In *Proc. of the 2nd HLRS Parallel Tools Workshop*, pages 157–167, Stuttgart, Germany, July 2008. Springer. ISBN 978-3-540-68561-6.

Chapter 8
Scalable Parallel Debugging with g-Eclipse

Thomas Köckerbauer, Christof Klausecker, and Dieter Kranzlmüller

Abstract Simulation software on today's HPC systems needs to be scalable to a large number of processes to make efficient use of such machines. Already at this level and especially with the expected increasing scalability of upcoming machines, the development and debugging of parallel programs becomes an increasingly difficult task. Consequently, sophisticated tools providing mechanisms for handling large-scale parallel and distributed programs are needed. In this paper we show several ways to improve the handling of large event traces using the Trace Viewer plug-in of the g-Eclipse tool and we propose the use of a pattern matching technique to simplify the debugging of large message passing parallel programs. With the pattern matching approach, we enable an additional layer of abstraction, which supports the user in understanding the program's behaviour.

8.1 Introduction

Locating errors in defective programs is a time consuming and difficult process. This is even more true for concurrent programs, due to additional error sources introduced by their parallel nature. Simplified, we can identify the following three reasons why debugging parallel applications differs from debugging sequential applications [5]:

- Increased complexity
- Amount of debugging data
- Additional anomalous effects (e.g. race conditions)

Thomas Köckerbauer, Christof Klausecker, Dieter Kranzlmüller
MNM-Team, Ludwig-Maximilians-Universität München (LMU)
Oettingenstraße 67, 80538 Munich, Germany
e-mail: koecker@nm.ifi.lmu.de

M.S. Müller et al. (eds.), *Tools for High Performance Computing 2009*,
DOI 10.1007/978-3-642-11261-4_8, © Springer-Verlag Berlin Heidelberg 2010

Due to the increased complexity of parallel and distributed programs it is hard to find errors by relying only on textual debugging information. For this reason, sophisticated program analysis tools supporting the developer during the debugging task by raising the level of abstraction from the textual representation of the source code are necessary. Even though a variety of such tools is already available, it is still questionable how well today's tools deal with the increasing number of processors in parallel systems, since the amount of debugging data is growing with them. In this paper, we discuss a possible solution to this problem with the Trace Viewer, a component of g-Eclipse, using several abstraction techniques. Furthermore, we provide some details about the included pattern matching technique, which identifies pre-defined communication patterns in event traces.

In [1] we provide an overview of the parallel application debugging capabilities of g-Eclipse [3][1 2]. g-Eclipse is an integrated workbench framework allowing to access the power of diverse computing infrastructures. It provides tools for Grids, Clouds, and High Performance Computing covering user, operator, and developer tasks. g-Eclipse is built on top of the Eclipse framework and continues as an Eclipse Technology Project[3] released under the Eclipse Public License (EPL). Currently version 1.0 of g-Eclipse is available for download.

8.2 Related Work

Due to the complexity of parallel programs and the resulting need for tool support, a broad spectrum of tools has been created, helping the developer to analyse and debug such programs.

Besides tools that search for potential problems without the collection of trace data, for example by performing run-time correctness checks like Marmot [4], also tools based on recorded program traces are available. The traces can be used to generate a communication graph enabling post-mortem program analysis.

One of those graph visualization tools is the Trace Viewer plug-in [1] which is integrated into g-Eclipse. It is based on the experience gathered with ATEMPT [6] and DeWiz [7] and can be seen as their evolutionary successor.

VAMPIR [11] is another well-known tool using trace data for parallel program analysis. Its main focus is on performance optimisation of parallel programs, whereas the Trace Viewer is targeted at supporting the debugging process. However, the basic data used by VAMPIR and the g-Eclipse Trace Viewer is compatible, and therefore the techniques could be used in both tools.

With the Trace Viewer, we pursue the approach of applying pattern matching techniques on trace data to find relevant structures in communication graphs enabling to enhance the graphical representation by highlighting the found structures.

[1] g-Eclipse has been partially supported under the European Union's 6th Framework Programme from 07/2006 until 12/2008, Contract Number IST-0343272.

[2] `http://www.geclipse.eu/`

[3] `http://www.eclipse.org/geclipse/`

In related works, different aspects of using pattern matching to analyse message passing programs have been employed.

Martino et al. proposed the use of static analysis in combination with solving diophantine sets of inequalities to find patterns in parallel programs on source code level [10]. This approach depends on the possibility to evaluate conditions influencing sent messages without actually running the program. The communication depending on information only present at run-time can not be analysed this way.

A different approach is to use suffix trees to find repeating communication events in traces [12]. One of the issues of this approach is to determine which sequence of repeating events to identify as the actual communication pattern, since the detected instances are typically not well-known communication structures but concatenations of them. While the technique is useful to find potential patterns in programs even if the expected communication structure is unknown, it does not allow to identify the kind of the detected patterns.

Ma et al. use pattern matching on graphs containing one node per process, and edges between the nodes communicating during the program execution [9]. This allows to identify communication patterns in programs not containing more than one pattern, or in selected parts of a program execution. However, it does not allow a distinction between different patterns with identical structure.

One of the many applications of pattern matching is the search for inefficient communication structures, with the goal of replacing them with more efficient ones and thereby improving the overall performance. This can be achieved by steering source code transformation using the information gathered from analysing trace data [13].

Another Eclipse project focused on providing tool support for parallel program development and debugging is the Parallel Tools Platform[4] (PTP). While PTP also focuses on scalable debugging of parallel programs, it does not include tools to analyse trace data, whereas we want to approach the debugging problem by using scalable trace visualization.

8.3 Debugging Parallel Programs Using g-Eclipse

The HPC specific program development and debugging functionality of g-Eclipse consists of three major components:

- Trace Viewer
- Remote Builder
- Remote Application Launchers

The central element is the Trace Viewer component, a tool allowing to visualize and analyse the communication of parallel message passing programs. It has been

[4] `http://eclipse.org/ptp/`

designed to be extensible and with future use in mind, enabling to contribute functionality via the Eclipse plug-in system. There are already several plug-ins available making use of the provided extension points.

The two existing trace provider plug-ins allow to open pre-recorded event traces in file formats from NOndeterministic Program Evaluator (NOPE) [8] and Open Trace Format (OTF) [2]. The visualization plug-ins use the provided data and take care of graphically presenting it. The provided core visualization plug-ins are based on the event-graph model, and enable to depict events in a space-time diagram according to their physical or logical timestamps. Additionally, a plug-in for displaying statistical data is available. The presentation of trace data can be altered by action and marker plug-ins, allowing to perform actions on, and change the appearance of certain parts of the event graph. Thereby, it is possible to highlight, hide, or re-order selected events and processes in the graph of a parallel program.

Apart from the Trace Viewer, the provided program development and debugging components include the Remote Builder, which allows to automatically build source code projects residing in the local g-Eclipse workspace on a remote machine. This enables the developer to compile applications for a remote environment without installing additional libraries, even for different architectures and operating systems.

After compilation, the generated executables can be started using the Application Launcher provided by g-Eclipse. The launcher enables to run and debug applications remotely, and also supports debugging of parallel programs comprising multiple processes.

To improve the debugging experience, the standard Eclipse debugging perspective was extended to ease the work of the developer when debugging parallel applications. This was achieved through integrating the Trace Viewer and establishing the link between trace visualization and symbolic debugging by enabling visual breakpointing, and by connecting graph and source code information [1].

Although program traces can easily occupy several hundred megabytes of storage, it is no problem to display such traces using the Trace Viewer, since all data is held in memory mapped files, mainly requiring virtual address space. However, while graphs with a small number of processes, or little communication, can be analysed by manual inspection, the presentation of programs with extensive interaction can get rather confusing, and it becomes hard for the developer to extract relevant information manually.

Therefore, ways to preprocess traces by filtering unnecessary or redundant data, and ways to identify important information in order to be highlighted in the event graph need to be established.

8.4 Reduction of the Trace Complexity

Since large traces contain an overwhelming amount of data, the information presented to the developer has to be filtered. For this reason the Trace Viewer contains several, manual and automatic, abstraction techniques which are integrated via the

extension points provided. This allows to reduce the amount of displayed information, hides irrelevant data or highlights events of interest by marking them in the event graph.
The provided trace filtering and highlighting features include:

- Grouping of processes in the trace using various techniques. This can, for instance, be useful to reduce the display of the trace by a dimension in the communication structure.
- Removal of all processes that are not involved in the communication with the process(es) of interest. This allows to only focus on communication that is neighbouring the processes in the communication structure (see Figure 8.1).

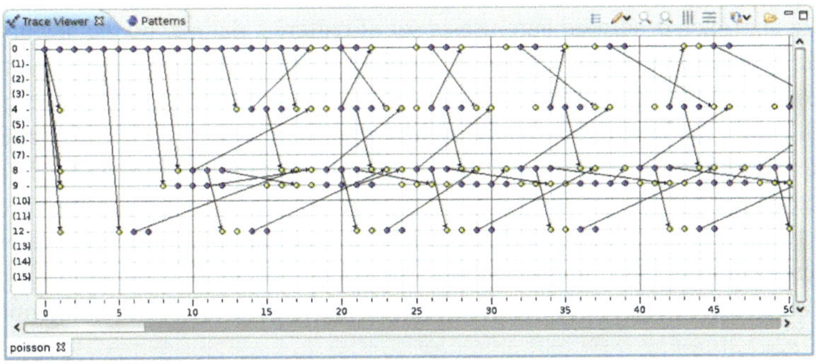

Fig. 8.1: Screenshot of a trace in which only the events of the selected process and its partners are displayed.

- Detection and marking of potential race conditions caused by wildcard receives. This enables to detect possible sources of non-determinism in parallel programs and allows to more easily compare different program runs.
- Highlighting of messages as well as events by matching user specified criteria, like for example the MPI routine, with event properties. Additionally, it is possible to unveil communication with different accepted and expected message sizes.
- Marking of patterns matched with a user defined set of pattern descriptions. This feature allows the distinction of parts in the trace that are composed from well known communication patterns from parts that do not follow those patterns.

8.5 Pattern Matching[5]

Traces of typical programs often contain repeating communication which follows certain patterns. To provide an insight into these communication structures, especially of large traces, different pattern matching techniques were developed.

In [13] an algorithm allowing to automatically extract repeating communication patterns from MPI traces using suffix trees was introduced. This helps providing a high-level understanding of the application's communication behaviour and allows to spot errors in communication, like an irregularity in the pattern.

Often it is the case that the expected communication structure in a program is known by the developer, and it would be beneficial for the developer to check if those patterns are appearing in an expected way.

The approach described here uses a set of patterns, defined by the developer and stored in a pattern database, which should be contained in the event graph. The pattern matching plug-in of the Trace Viewer allows to filter well known patterns, which can be defined using a pattern description language. The found patterns are marked in the trace to distinguish them from the rest of the events. Since the displayed graph in the Trace Viewer contains marked patterns which can be recognised quickly it is easier to focus on communication that does not follow the well known, expected patterns. The unmarked events may be the result of an error that interfered with the communication in a way that it does not follow the patterns anymore.

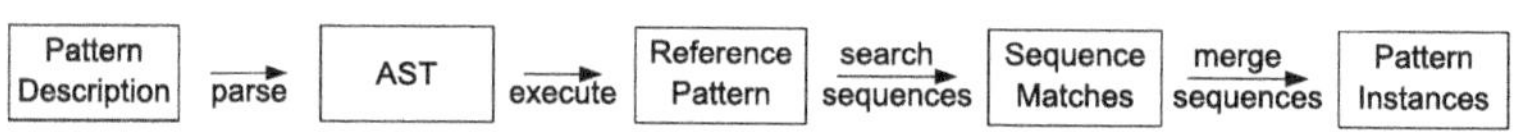

Fig. 8.2: Steps of the pattern matching process.

The pattern detection in the proposed approach is done in several steps (see Figure 8.2):

1. The pattern descriptions, created by the developer, are parsed and abstract syntax trees (AST) of the descriptions get built.
2. The ASTs of the descriptions get executed for the desired process count, thereby producing a reference pattern matching the scale of the trace to search in.
3. These reference patterns are stored on disk, since they can be reused if there is another search for this pattern in a trace with the same amount of processes. In this case (provided that the pattern description was not changed in the meantime) the steps 1 and 2 are skipped in the next search.
4. A hash-based search on the individual processes of the trace is done to find potential matches on a per-process base. This search also considers variations of the patterns, which can for example be caused by a reordering of the process IDs in comparison to the reference pattern.

[5] Section 5 on pattern matching is part of the PhD thesis of Thomas Köckerbauer

5. The potential matches on the different processes are merged and get checked if they form a described pattern.

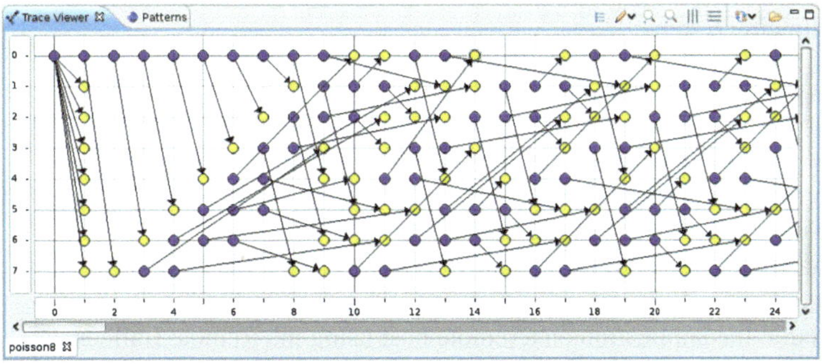

Fig. 8.3: Screenshot of a trace without pattern marker. The send and receive events are depicted in different colours.

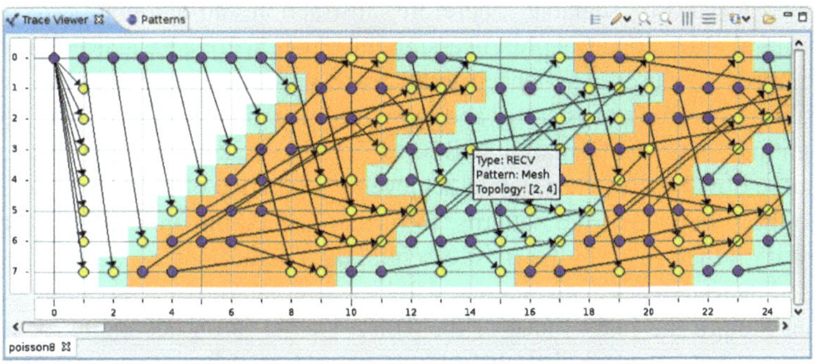

Fig. 8.4: Screenshot of a trace with pattern marker. The tooltip shows the event type and the identified pattern instance.

Figure 8.3 shows a trace with only a few processes and although it is easy to see that there is some repeating pattern it already requires some time to find out which pattern it contains. Figure 8.4 shows the same trace with the pattern marker applied which highlights the found pattern instances. The tooltips of the events in the pattern now additionally contain information about the found pattern. In the example of Figure 8.4 it shows that the detected pattern is a "mesh" pattern with the dimension 2x4.

8.6 Summary and Conclusions

In this paper we suggest an additional step to be inserted into our previously established debugging approach. The resulting debugging approach can be divided into the following 6 steps:

1. Record the trace data during an initial program run.
2. Apply abstraction mechanisms provided by the Trace Viewer to identify errors.
3. Set breakpoints on events in the event graph using the Trace Viewer.
4. Launch applications remotely with debuggers attached to the individual processes using an Application Launcher.
5. Use Eclipse's standard debugging features to debug the remotely running parallel applications.
6. Rebuild the application on a remote host using the Remote Builder.

In this paper, we have shown how using event graph visualization instead of relying only on textual representation of information can benefit the debugging process. We have introduced several abstraction mechanisms allowing to use the event graph even for complex applications comprising a large number of processes and extensive interaction. With the prospect of even more processing cores in future parallel systems, debugging and analysis software tools need to be flexible, extensible, and - first and foremost - scalable. Therefore, abstraction and automation techniques, supporting the developer by preprocessing information are a basic necessity.

To improve the user experience for the Trace Viewer even more, and to make the accessibility of large traces easier and more intuitive, other means of extracting and highlighting relevant information will be explored to make most significant correlations more obvious. Another step to achieve the objective of scalable trace analysis will be the grouping of the patterns, to further reduce the amount of information displayed on the screen to the most essential data. Additionally, ways to refine the proposed pattern matching technique will be investigated.

Even though the main focus of the Trace Viewer is on debugging of message passing programs we are also planning on extending it by integrating support for performance analysis tools in order to make our tool even more versatile.

References

1. Christof Klausecker, Thomas Köckerbauer, Robert Preissl, and Dieter Kranzlmüller. Debugging MPI programs on the grid using g-Eclipse. In Michael Resch, Rainer Keller, Valentin Himmler, Bettina Krammer, and Alexander Schulz, editors, *Tools for High Performance Computing, Proceedings of the 2nd International Workshop on Parallel Tools for High Performance Computing*, pages 35–45, Stuttgart, July 2008. HLRS, Springer-Verlag.
2. Andreas Knüpfer, Ronny Brendel, Holger Brunst, Hartmut Mix, and Wolfgang E. Nagel. Introducing the open trace format (OTF). In *ICCS 2006*, volume 3992/2006 of *Lecture Notes in Computer Science*, pages 526–533. Springer Berlin / Heidelberg, 2006.

3. Harald Kornmayer, Mathias Stümpert, Markus Knauer, and Pawel Wolniewicz. g-Eclipse - an integrated workbench tool for grid application users, grid operators and grid application developers. In *Cracow Grid Workshop '06*, Cracow, Poland, October 2006.
4. Bettina Krammer, Katrin Bidmon, Matthias S. Müller, and Michael M. Resch. MARMOT: An MPI analysis and checking tool. In *ParCo*, pages 493–500, 2003.
5. Dieter Kranzlmüller. *Event Graph Analysis for Debugging Massively Parallel Programs*. PhD thesis, Johannes Kepler University Linz, September 2000.
6. Dieter Kranzlmüller, Siegfried Grabner, and Jens Volkert. Event graph visualization for debugging large applications. In *SPDT '96: Proceedings of the SIGMETRICS symposium on Parallel and distributed tools*, pages 108–117, New York, NY, USA, 1996. ACM.
7. Dieter Kranzlmüller, Michael Scarpa, and Jens Volkert. DeWiz - a modular tool architecture for parallel program analysis. In *Euro-Par*, pages 74–80, 2003.
8. Dieter Kranzlmüller and Jens Volkert. NOPE: A nondeterministic program evaluator. In *ParNum '99: Proceedings of the 4th International ACPC Conference Including Special Tracks on Parallel Numerics and Parallel Computing in Image Processing, Video Processing, and Multimedia*, pages 490–499, London, UK, 1999. Springer-Verlag.
9. Chao Ma, Yong Meng Teo, Verdi March, Naixue Xiong, Ioana Romelia Pop, Yan Xiang He, and Simon See. An approach for matching communication patterns in parallel applications. In *IPDPS 2009*, 2009.
10. Beniamino Di Martino, Antonino Mazzeo, Nicola Mazzocca, and Umberto Villano. Parallel program analysis and restructuring by detection of point-to-point interaction patterns and their transformation into collective communication constructs. *Science of Computer Programming*, 40(2-3):235–261, 2001.
11. Wolfgang E. Nagel, Alfred Arnold, Michael Weber, Hans-Christian Hoppe, and Karl Solchenbach. VAMPIR: Visualization and analysis of MPI resources. *Supercomputer*, 12(1):69–80, Jan 1996.
12. Robert Preissl, Thomas Köckerbauer, Martin Schulz, Dieter Kranzlmüller, Bronis R. de Supinski, and Daniel J. Quinlan. Detecting patterns in MPI communication traces. *International Conference on Parallel Processing*, 0:230–237, 2008.
13. Robert Preissl, Martin Schulz, Dieter Kranzlmüller, Bronis R. de Supinski, and Daniel J. Quinlan. Using MPI communication patterns to guide source code transformations. In Marian Bubak, G. Dick van Albada, Jack Dongarra, and Peter M. A. Sloot, editors, *ICCS (3)*, volume 5103 of *Lecture Notes in Computer Science*, pages 253–260. Springer, 2008.

Chapter 9
New Analysis Techniques in the CEPBA-Tools Environment

Jesus Labarta

Abstract The CEPBA tools environment is a performance analysis environment that initially focused on trace visualization and analysis. Current development efforts try to go beyond the presentation of simple statistics by introducing more intelligence in the analysis of the raw data.
The paper presents an overview of three recent developments in this area. First, we show how spectral analysis techniques can be used to isolate sufficiently small regions of a trace that characterize the behavior of the whole run. Second, we describe how clustering analysis techniques can be used to identify temporal and spatial structure in parallel programs, an essential component to ease the job of the analyst, but also to automatically derive a broad range of both precise and focused metrics from a single run of a program. Then we describe how sampling and tracing data acquisition techniques can interoperate to generate with very low overhead extremely precise metrics about the temporal behavior of a program.
The development rests upon the trace based CEPBA-Tools environment, using the Paraver visualization capabilities to check the quality and usefulness of the techniques. Once identified, they can be implemented on-line aiming at maximizing the amount of information obtained from a run. We report the work being done on top of MRNET in this direction.
We consider that by applying and combining these and other techniques from various data analysis and mining fields, performance analysis tools will be able to effectively address the huge challenge posed by future exascale systems.

Jesus Labarta
Barcelona Supercomputing Center and Technical University of Catalonia,
Jordi Girona 29, Barcelona, Spain,
e-mail: `jesus.labarta@bsc.es`

M.S. Müller et al. (eds.), *Tools for High Performance Computing 2009*,
DOI 10.1007/978-3-642-11261-4_9, © Springer-Verlag Berlin Heidelberg 2010

9.1 Introduction

As larger and larger systems are being developed and applications run on them, the issue of understanding how they behave and how efficiently our applications use the available resources is more and more important.

Performance analysis tools rely on hooks injected into programs to capture relevant events and derive the metrics that quantify and explain their behavior from the acquired data. Traditionally the focus of performance analysis tools has been centered on the monitoring or data acquisition mechanisms. The algorithms used for processing the raw data before presenting results to the analyst are typically very simple. Profilers are the more widely used type of tools and they just present simple statistics like time in each routine, total count of invocations or the accumulated instructions. By aggregating over the time and processor dimensions and focusing on a limited set of predefined metrics, profilers reduce the amount of data that has to be emitted and then presented to the user. This advantage comes at the expense of loosing detail on the variability of system activity and results in a lot of relevant information being discarded.

Intermediate approaches with different amounts of precomputed profile data have been used but the question arises as to how raw data should be processed to maximize the relevant information obtained from it while minimizing the amount of data emitted.

Other areas of science and engineering have developed elaborated techniques to extract useful information out of the raw data. Signal and image processing and data mining techniques are widely used in different fields with such purpose. We have the perception that performance analysis lags far behind other areas in the actual use of those techniques as well as the conviction that they could be successfully used in our field.

In this paper we describe some of the usages of signal processing and data analysis techniques within the CEPBA-tools environment. Section 9.2 briefly describes the environment used to develop and validate the approaches described in successive sections. Section 9.3 then focuses on the use of spectral analysis techniques, section 9.4 on the use of clustering techniques and section 9.5 on the combined use of instrumentation and sampling. Section 9.6 describes current work in integrating the above described techniques in an automatic on-line analysis environment and section 9.7 presents some views on future directions.

9.2 The CEPBA-Tools Environment

The development of the CEPBA-tools environment started in 1996 [5] with the objective to better understand the detailed interactions that could take place in a multiprogrammed message passing machine based on the Transputer chip. Three main components constituted the environment: a set of tracing packages (now MPITrace) for message passing programs, a simulator (Dimemas) for message passing machines also modeling the time sharing behavior within a node and a visualizer (Paraver) capable of displaying traces produced by the simulator. The traces capturing the actual behavior of a run of the parallel program could also be directly generated by the instrumentation package and visualized with Paraver. The usage of the tools then evolved to support detailed analysis of single applications and prediction of the impact of different architectural parameters in their performance.

Paraver is a flexible browser for traces that contain sequences of timestamped records of three types: events, states and communication. The Paraver trace format describes the structure of these records but their semantic is essentially undefined, which gives the possibility to apply the tool in very different areas, areas far beyond those initially targeted. This is certainly the case for the records that represent a punctual event with two attributes (type and value) and for state records that represent an interval between start and end for which one attribute is given. The attributes are integer values in which the tracing package can encode the information as desired. Each record applies to one object in a hierarchical structure of three levels which when instrumenting parallel programs are typically mapped to application, process and thread. Communication records actually relate two such objects and have two additional attributes.

The core of the Paraver engine [12] is called the *semantic module*. It provides through its GUI a very flexible algebra to specify how functions of time can be generated out of the records and the numerical values of their attributes. One such function of time is generated for each object. The fact that internally Paraver considers the data it handles as functions of time leads naturally to some of the techniques described through the paper. Finally, a simple but flexible rendering mechanism translates the functions of time to colored timeline plots. Typically a palette of colors is used to translate categorical valued functions of time such as identifier of the MPI call, or user function. A gradient color map is used for continuous valued functions, using light green for low values up to dark blue for large values. Areas where the function value is above a specified range are highlighted in orange and if the function value is zero, the background color is used. Non linear rendering is used to expose information to the analyst in cases where many values map to a pixel. This technique addresses the scalability issues faced when displaying traces with many objects or representing long time intervals.

The *analysis module* implements a single mechanism to compute tables. A very generic approach is used, able to not only report statistics but also histograms and correlations between any of the functions generated by the *semantic module*.

Complex expressions can be defined in the *semantic* and *analysis modules* and saved along with the display setup in configuration files for later reuse. The lack of semantics in the trace format plus the flexibility of these two modules makes Paraver an extremely powerful and versatile browser. It has been used to analyze MPI and MPI+OpenMP programs but also operating system activity, multicore architectures, or file system behavior. Other time series not having any relationship to parallel programming such as stock sensor data or exchange rates could be analyzed in Paraver without requiring any modification of its source code and without requiring convoluted mappings of concepts in these areas to the concepts handled by visualizers too specialized in just parallel program.

9.3 Spectral Analysis

Many applications tend to have an iterative structure, originating from the time stepping process they often simulate. The behavior of such iterations tends to be very repetitive or at most slowly varying as the simulated system evolves. This means that a few iterations are sufficient to describe the behavior of applications during long intervals of time. Spectral analysis techniques can be used to determine the periodic structure of a program. One of the applications of such analysis is to automatically select the time interval to be traced such that at least one whole period is captured.

Other sources of repetitive behavior are the iterative nature of the numerical algorithms, the need to process a large number of particles or elements, and so on. These iterative patterns may be nested but for a global performance analysis purpose we are mostly interested in the outermost levels. In [4] we showed how the iterative behavior at different levels can be identified on traces from large runs of a program. The main usage in that work addressed the possibility of reducing the size of the traces required to still be able to do very detailed analyses.

The spectral analysis can be applied to signals representing the evolution with time of some metric for the whole applications, such as average instantaneous instructions per cycle (IPC), or actual number of processes inside MPI calls. The paper also revealed that it is not necessary to use signals representing a metric meaningful from the performance point of view. In fact, the sum at each point in time of the duration of the computation burst of all the processes is a signal with no real meaning that captures pretty well the structure of an application. A computation burst is the time interval between exit of an MPI call and entry to the next. During the whole burst, a process contributes with its duration to the global function. While a process is inside MPI no contribution is made to the global signal.

In the same study we also showed how other techniques such as mathematical morphology can be used to clean-up signals. This non linear filtering technique was applied to signals identifying regions where certain type of perturbations occurred while obtaining the trace. One example of such a signal is the number of processes flushing their trace buffer to disk. Although with sufficiently large buffers this will not be very frequent, it will certainly perturb not only the process doing the flush but also other processes communicating with it. Furthermore, it is frequent that different processes flush their buffers at about the same time. By applying dilation and erosion filters to such signal it is possible to separate regions in the influence area of the perturbations from large regions without such perturbations.

Other technique to obtain general structural information of the trace is the Wavelet transform. This can be used to automatically separate the non iterative phases of an application such as initialization and termination from the core computation phase. The Wavelet transform produces information about the spatial localization of energy at different frequencies. When applied to signals like the sum of the useful duration described previously, initialization and termination phases tend to have much lower energy at high frequencies. By applying again mathematical morphology techniques to the high frequency outcome of the wavelet transform we can identify regions of major program activity.

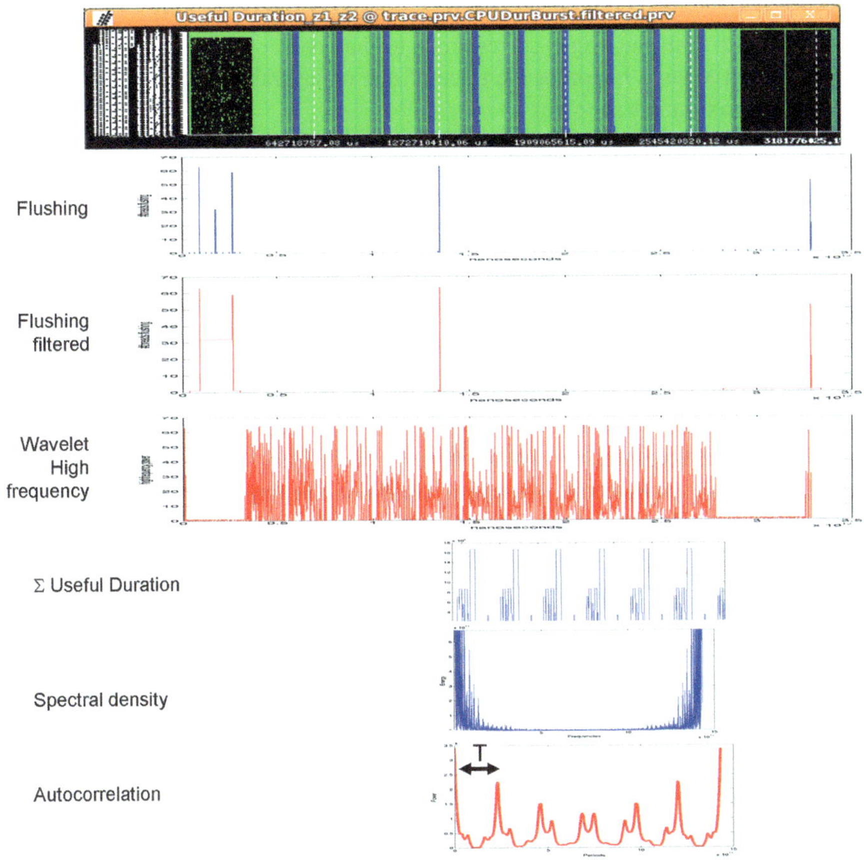

Fig. 9.1: Process of automatic period detection

The whole process is described in figure 9.1. The timeline on top represents the duration of the computation bursts for each of the 128 processes of a run of the WRF weather modeling code. Dark blue represents long computation bursts, light green short computation bursts and black corresponds to time inside MPI. The two signals below represent when some process is flushing data to disk. At the scale shown there is no appreciable difference between them, but the second one corresponds to the outcome of the filtered signal with mathematical morphology. A look at a more detailed scale shows that several flushes from different processes have actually been merged into a single burst. The fourth view from top represents the high frequency components identified by the Wavelet transform when applied to the useful duration signal. The main computation area corresponds to the region with high values. Combining this signal and the outcome of the flush analysis the tool identifies the longest core computation region without perturbation and builds the useful duration signal for that interval. Computing the FFT, squaring it and computing the inverse we obtain the autocorrelation function shown at the bottom of the figure. Peaks in this figure correspond to periodicities in the signal. In our case, the first local maxima different from the origin corresponds to the coarser periodicity. The tool can then be used to cut a region of the trace of one or several periods (depending on a requested maximum trace size).

This functionality was developed as a command line tool to process large traces and is now being integrated both in the Paraver GUI and in the intelligent on-line tracing packages as described in section 9.6.

9.4 Clustering Techniques

Clustering techniques have been used in the parallel performance analysis area mostly with the aim of identifying groups of processes of differentiated characteristics. The target has typically been to obtain a representative process for each group and thus reduce the number of processes on which to carry out further analyses.

In [2] we aimed at using clustering techniques with the objective of identifying internal structure at the level of computation bursts within the application. We try to group computation bursts between MPI calls by their similarities in terms of duration and hardware counter derived metrics. In the following sections we describe the relevant data processing and clustering algorithms, usage examples clustering and further work on automatic quantification of the quality of a clustering result.

9.4.1 Clustering Algorithms

Given our objective, the data to be clusterized corresponds to each of the computation bursts between MPI calls. For a typical trace there may be many thousands or millions of records, each of them characterized as a point in an N dimensional space. Possible dimensions include the duration of the region, the number of instructions, cache misses or other captured hardware counters. The number of these dimensions is limited by the number of hardware counters that can be simultaneously read, but derived metrics between several hardware counters such as IPC on miss ratios can also be used.

In order to keep the clustering algorithm in reasonable times and to focus the efforts in relevant regions we filter out bursts that are either very short or have a value of some of the counter below a threshold. The user can specify through an xml file these different thresholds as well as the metrics to consider as dimensions in the clustering algorithm, further data preprocessing transformations (ie. scaling, principal components,...) or other parameters required by the algorithm.

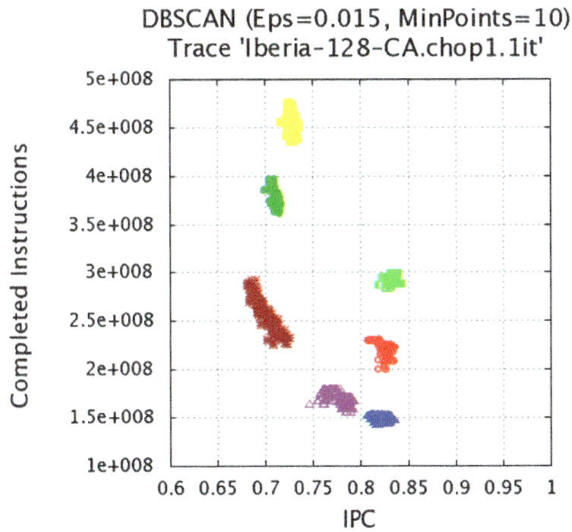

Fig. 9.2: Scatter plot of clustered WRF bursts

We use DBSCAN, a density based algorithm, as we have observed that the assumption made by k-means type of algorithms that data distribution is spherical in nature does not hold with our data. Figure 9.2 shows an example projection of points of a weather forecast run (WRF) on the Instruction and IPC dimensions. We can see how some clusters do have a spherical shape with little variability in both dimensions. Others show a negative correlation between instructions and IPC: the larger the instruction count in the bursts the lower the IPC. The reverse situation may hold on other cases. Clusters where the same number of instructions are executed with a wide range of IPCs are also frequent.

9.4.2 Application of Clustering Techniques

The presentation of the scatter plots such as the one in figure 9.2 does provide a lot of information to the analyst on the behavior of the different regions, but the doubt may arise as to how do the identified clusters distribute over time. The tool can inject new events into the original tracefile labeling each computation burst with its identified cluster. In this way it is possible to visualize the space and time distribution of the clusters. This conveys to the analyst complementary information to the scatter plots, reflecting in detail the structure of the application behavior. Figure 9.3 shows the cluster timeline corresponding to figure 9.2.

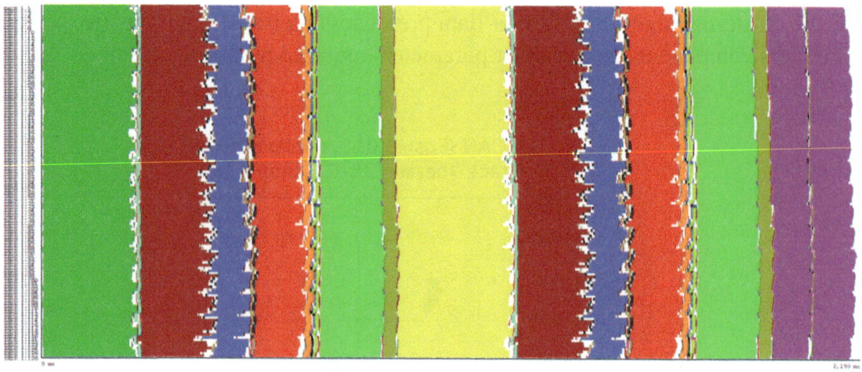

Fig. 9.3: Clustered WRF timeline

Deriving precise metrics and models of the performance of the sequential computation phases is another important use of the clustering techniques. Current processors do have the ability to perform very detailed counts of their internal activity and such information is made available through APIs, of which PAPI [10] is the most widely used. For cost reasons and although the list of potential counts is quite large, the actual number of counters that can be read at the same time is limited and architecture dependent. Being able to read many counters would also introduce significant overheads on the monitoring system. The result is that each data acquisition captures only partial information and in order to obtain values for a large set of counters, either several runs or sampling techniques have to be used. In these approaches, a lot of precision in the ability to correlate counts for individual regions of code is lost.

The use of the clustering techniques we have proposed provides an alternative to achieve higher precision still requiring only a single run of the program. The idea is to shift over time or processes the set of counters acquired with the constraint that a couple of them (typically cycles and instructions) be present in all sets. The acquisition does not even need to be synchronized across processes. The only requirement would be that a sufficiently large run is made such that several acquisitions with different hardware counter sets are made for the relevant computation bursts. If the application does have an SPMD structure, different sets can be used on different processors to reduced the required duration of the acquisition process. By clustering all the points using the two common counters, each point is assigned to one cluster and thus contributes its non common counters to the characterization of such cluster. The derived metrics computed from counts of different instances of a cluster are certainly approximations, but the accuracy is much higher than if the correlation is made through a blind statistical sampling process or derived from two independent runs.

An example use is show in figure 9.4. From a single run of the program, a very large number of hardware counters are obtained for the major clusters in the program. Those counters are fed into a cycles per instruction (CPI) stack model [11]

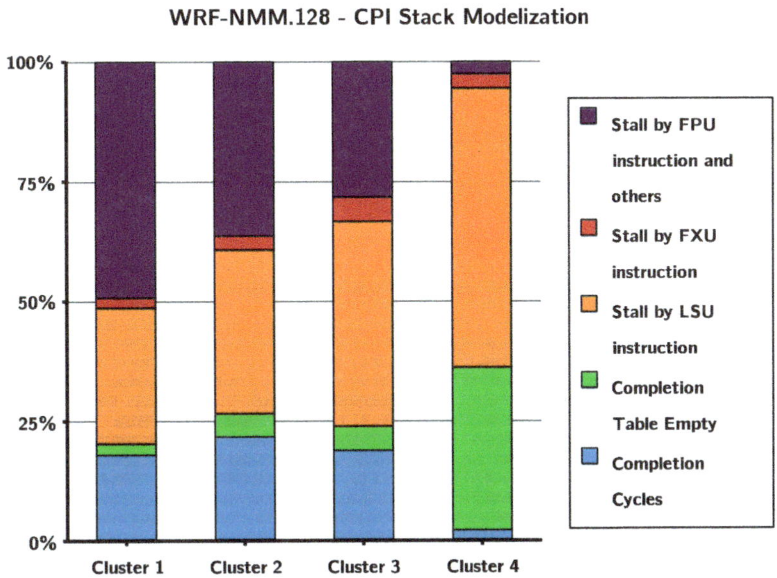

Fig. 9.4: CPI stack model derived from a single run

to obtain a fair quantitative description of which components in the architecture are determining the sequential computation performance of each cluster.

9.4.3 *Quantification of the Clustering Quality*

Given that most programs do have an SPMD structure, cluster timelines like the one shown above should display all processes as being in the same cluster at a given point in time. Of course some skew in the time each process enters a cluster is possible. Also different instances of the same cluster may be of different length in different processors, but in general, vertical stripes would be expected. When blindly applying the clustering algorithm it is nevertheless possible to obtain timelines like the one shown in figure 9.5. There we see that some parts do show an SPMD structure while in other intervals different processors are in different clusters. This is caused by real differences in the behavior of processors but the question is what level of granularity we want the clustering algorithm to use when determining what is similar and what different. A noisy plot with a lot of clusters, indicating that everything is different does not actually convey to the analyst useful information on the structure of the application. A plot with just one cluster, where everything is the same does not convey information either. The granularity used by the clustering algorithm is controlled by a parameter provided by the user. Although a wide range of values of such parameters give in general useful results for the DBSCAN algorithm, it is possible to tune it if coarser or finer detail is desired. Typically clustering result with a handful of clusters (less than 10), representing more than 90% of the total execution time and showing an SPMD structure could be considered as a good result for a first analysis.

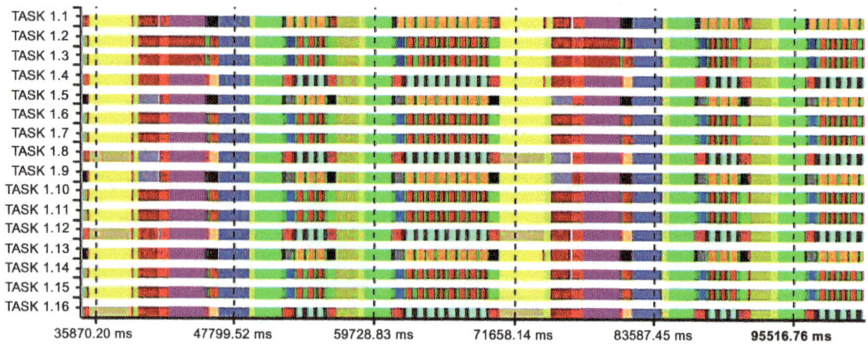

Fig. 9.5: Clustered timeline of WRF run on 16 processors

An interesting question is how the above criteria could be automatized. In [3] we presented an approach to automatically detect the *spmdiness* of a clustering result. The idea builds on the similarity between a process seen as a sequence of clusters and a sequence of aminoacids as handled in the life sciences area. In this field, a lot of tools have been developed to check the alignment and similarities between sequences of such chains. Our approach is to leverage that technology and use available Multiple Sequence Alignment (MSA) codes to properly align the different clusters executed by each process. In doing the transformation of the timeline to a sequence of clusters, the actual duration of the different instances is dropped and the focus shifts to their sequence, which in itself provides yet another way of representing the internal structure of the computation.

Figure 9.6 shows this situation for a run on 16 processors of the NAS LU benchmark (class A). On top we see the timeline of cluster while the bottom figure shows structure in terms of the aligned sequence of clusters as reported by the Kaling2 package. In this case we see that the time dimension has disappeared and each cluster uses only one column, irrespective of whether it was large or small. An asterisk on top of the image shows that the tool has been able to perfectly align that column. Lack of the asterisk means that there is either some hole in that column or not all the clusters are identical.

At the beginning of the timeline we observe several long clusters, with a quite good SPMD structure although task 6 is in a different cluster when all other processors are in the green cluster. In the bottom view of the figure we can see that this outlier causes the first non perfectly aligned column. Towards the end of the timeline each process executes a series of yellow clusters. It is unclear in such view whether all the processes have the same number of instances of the yellow cluster or not, and they are certainly skewed in time. The bottom figure shows how the alignment of those clusters is also quite good although some misaligned positions show up.

From the output of the alignment tool we can compute a metric specific to our purposes that quantifies with a number between 0 and 1 the *spmdiness* of each individual cluster as well as the global *spmdiness* of the clustering result. These metrics tell the analyst how good or bad the clustering result is and will in our future work be used as the quantification method to automatically search for a good granularity for the clustering algorithm. It will be possible to separate sets of points for which a coarse granularity results in good SPMD characterization from other sets where detailed granularity can extract finer detail and still reflect a good SMPD structure.

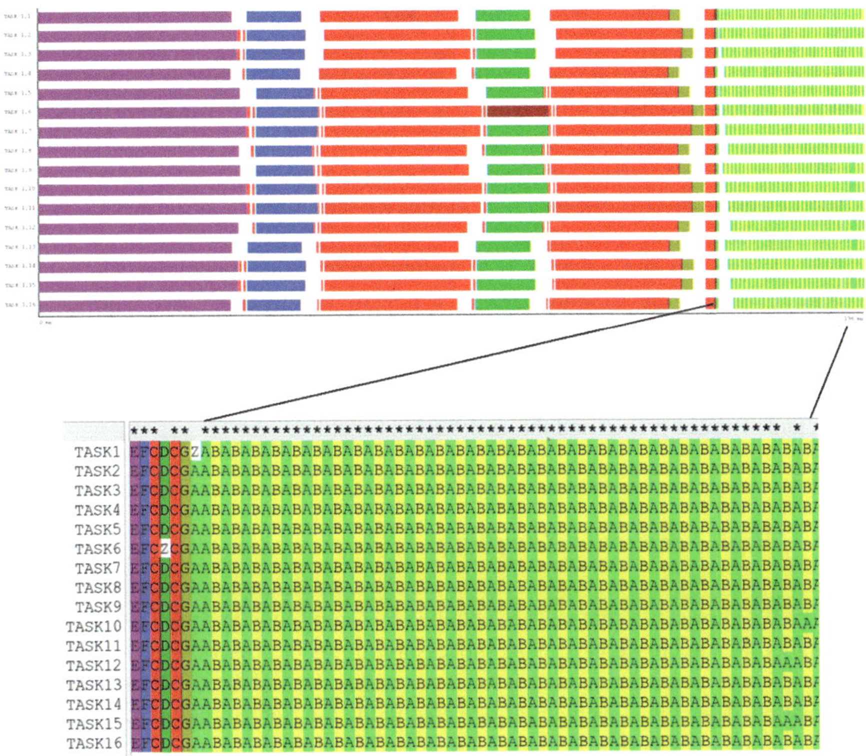

Fig. 9.6: Timeline and alignment of clusters for a 16 processor run of the NAS LU benchmark (class A)

9.5 Sampling and Mixed Instrumentation

Statistical sampling is a technique used by profile tools as a way to obtain approximate characterizations of a program without requiring to instrument neither the source nor binary. Sampling uses a mechanism external to the application control flow to fire a probe that from time to time captures information about the program such as the line in the code being executed. From these counts, we estimate the percentage of time spent in each of the lines by assuming it is proportional to such counts. A periodic sampling every few milliseconds is often used. Besides the time based approach, the firing of the probe can be based on the overflow of a hardware counter such as cache misses or floating point operations. In this case, the count for each line is expected to be proportional to the number of cache misses incurred or floating point operations performed by that line.

As mentioned before, the evolution of a program activity can be seen as a signal with spectral components corresponding to the speed of change in the activity of the program. As well known in signal theory, a function can be perfectly characterized by its samples if taken at sufficiently high frequency, in particular above the Nyquist frequency of the signal. In our context, this means that if the sampling frequency is sufficiently high compared to the rate of change in the application, we will be able to get a good picture of the evolution of metrics such as mips, cache misses, or other data captured during the sample.

Tools typically use either an instrumentation or a sampling approach, but typically not both at the same time. Even if they do, it is to obtain two types of information (ie. counts and percentage of time) which are then not correlated, at least as much as we consider it could be done. The following subsections show how the MPITrace package in CEPBA-tools has been extended to handle both instrumentation and sampled probes and the analyses this enables.

9.5.1 High Frequency Sampling

Figure 9.7 shows different metrics obtained by sampling with a period of 1M cycles (roughly 445 microseconds) a 16 processors run of the NAS BT benchmark. The top view represents the main user functions and is derived from instrumentation information. The color scheme is as depicted in the top right corner of the figure. The view can be used as time reference for those below and shows how they represent about 1.5 iterations of the program. One iteration takes in the order of 345 ms, thus the number of samples in it is around 760, a large enough number to capture a lot of detail.

The four views on the bottom show the evolution over time of four metrics: mips, load mix (percentage of load instructions with respect to total number of memory access instructions), memory mix (with respect to total number of instructions) and the ratio of L1 to L2 misses. The figure shows how routines x_solve, y_solve and z_solve have an initial phase with four steps of high mips, and low memory mix.

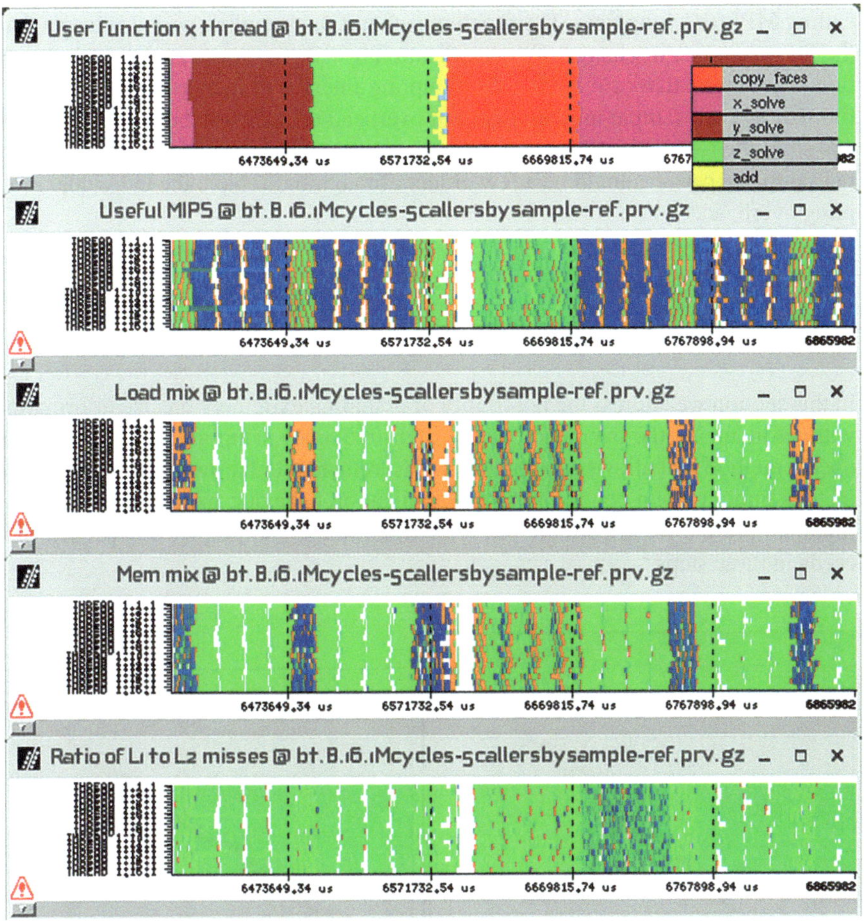

Fig. 9.7: Time evolution of metrics captured sampling at high frequency.Light green correspond to a low value of the metric, dark blue to a high value and orange to even higher values

It is interesting to see how routine x_solve has a higher L1 to L2 ratio of misses. Towards the end of each of these three routines, there is a phase where the memory mix, and in particular loads increases a lot, resulting in a lower mips rate.

In the above example, both instrumentation and sampling information are acquired by MPITrace, and they are visually correlated by the analyst. The instrumentation information can provide precise data about the structure of a program such as when a routine is entered and exited and how many instructions are executed in it. The granularity is nevertheless limited to the actual duration of these routines in the user code. The sampling information ensures a granularity (1M cycles in our example) at which data is acquired, even if the program stays within a routine without calling MPI for a long time. The problem is that if the sampling period is larger than the fine grain rate of change of the application it will not produce relevant information. The alternative of reducing the sampling interval to increase the precision is limited, as each sample implies a certain overhead to interrupt the process, capture and store the required information. In order to maintain the total overhead bounded, the sampling period should be a few orders of magnitude above the individual sampling overhead.

9.5.2 Hybrid Instrumentation and Sampling

In this section we address the possibility of obtaining extremely precise information without incurring the overhead of very high frequency sampling. In [1] we developed a method that allows such precise measurements for hardware counter derived metrics under certain conditions in the application behavior, namely the ergodicity (maintaining the same periodic behavior over time). The proposed approach proceeds in three steps.

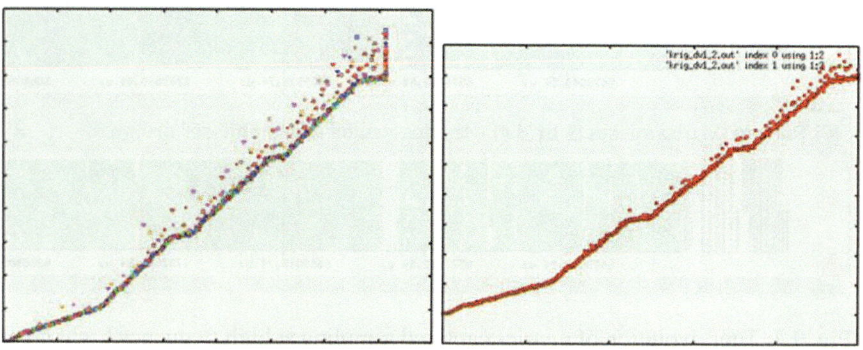

Fig. 9.8: Cumulative instruction count since the beginning of the iteration after the merging process. The left view includes all iteration instances. The right view show the detail when outliers have been discarded.

First we transform the captured hardware counts, which by default correspond to the value since the previous probe where the counters were read. This transformation is done by referring the counts at each sampled point to their nearest previous instrumentation event. By this we mean that we associate with such events the aggregated hardware count (instructions, cache misses,. . .) of all the previous probes since the reference. The same aggregated value is computed for the instrumentation event at the end of the region (iteration or routine). In this way we obtain for each instances of a region a list of monotonically increasing hardware counts timestamped with respect to the start of the region. The number of such points for one instance may be just two (entry and exit) if no sample fell inside the region or a very large number if the sampling period is much smaller than the duration of the region. In any case, the precision of this data is limited by the sampling period.

The next step is to merge the different instances in order to increase the density of points and thus the precision. If all instances took exactly the same amount of time, just merging and sorting the different lists by the timestamp of the entries would be enough. As this situation will never happen in a real system, we take one instance as reference and scale the timestamps of the lists of all other instances such that the duration of the region matches the reference. After scaling we apply the merge process just described. If the variance in the duration of the different instances is not high, this should result in a thin cloud of points around the actual cumulative distribution of hardware counts since the start of the region. The upper plot in figure 9.8 shows the result of this process for one iteration of one process in the NAS BT benchmark class A run on 16 processes. We can see a certain amount of variability. In order to reduce it, we try to identify region instances that are outliers in terms of total duration and do not merge their points into the final list. The result is shown at the bottom of figure 9.8, with significantly less variability. More restrictive selection of outlyers would reduce variability but also the amount of points and thus precision. Finally, the region can be analytically characterized by a polynomial fit of the cloud of cumulative counts. We tried different fitting models but finally decided for a Kringing method. The analytical expression can then be reported as output of the analysis process. It can also be sampled at periodic intervals and synthetic events injected in the trace. It is also interesting that derivatives can be computed, thus reporting instantaneous rates such as mips or flops.

By using these hybrid sampling and instrumentation techniques, it is possible to compare the instantaneous evolution of metrics reporting how well different parts of the core architecture are being exercised. By correlating them we gain a deep understanding of the behavior of a program. Figure 9.9 shows the normalized instantaneous evolution (from top to bottom and left to right) of mips, store mix, L1 misses and load mix. We can clearly identify an initial phase where the mips rate is high and has a high proportion of store instructions but very low L1 miss rate. After that, four iterations of phases with a fair mips rate are separated by transitions with low mips. Such low performance in the transitions is correlated to a high L1 miss rate also caused by a high proportion of load instructions.

We have described the process to obtain the instantaneous evolution of metrics derived from hardware counter reads. It is also possible to apply the folding process to the call stack information captured by the samples to obtain a timeline of the code line being executed along time. In this case it is not possible to perform an analytical fit of the time function and variability between different iterations will introduce a certain degree of inaccuracy (i.e. backwards control flow). The result is nevertheless extremely useful as shown in figure 9.10. We can identify the four iterations of an outer loop and see how the execution progresses through the code with some source lines making longer contributions to the execution time than others.

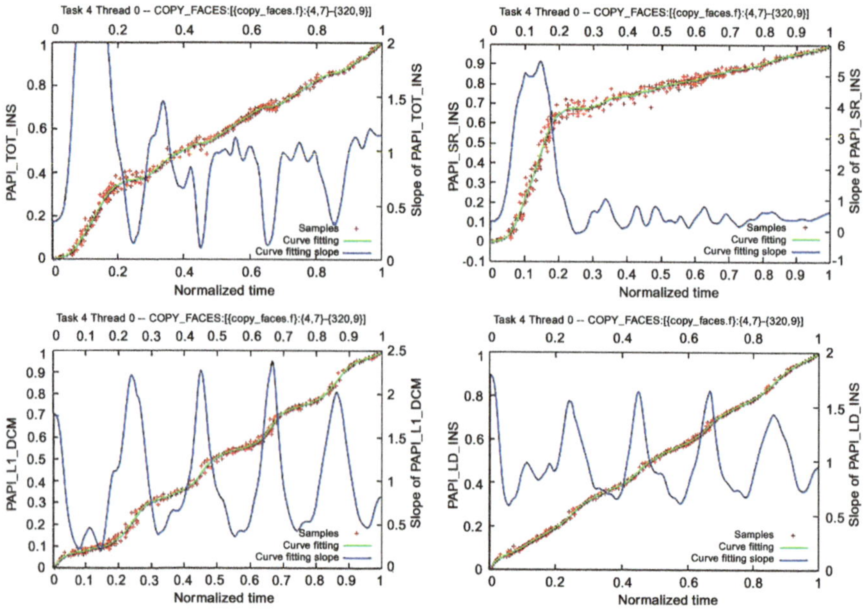

Fig. 9.9: Instantaneous evolution of different metrics for the copy_faces routine in the NAS BT benchmark

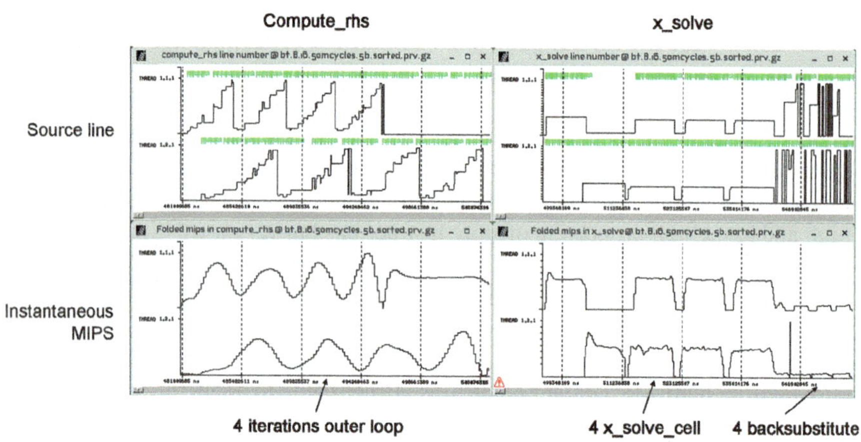

Fig. 9.10: Correlation between metrics and folded source line for two processes in a 16 processors run of the NAS BT becnhmark

9.6 On-line Techniques

The research activities described in the previous sections heavily used Paraver to visualize traces and validate the results of the different techniques. The question then arises whether those techniques can be applied on-line, to automatically summarize the data captured by the monitoring system and minimize the size of files it generates while maximizing the amount of information emitted.

In [7] we describe ongoing work integrating those techniques into our instrumentation packages and using MRNET [9] as a scalable infrastructure. Such on-line integration does pose new challenges and the need to adapt the basic algorithms accordingly. Two of these extensions are described in [7] and will be summarized in the next paragraphs. The first one addresses the issue of the overhead of the clustering algorithm itself. The second one looks at the stability of the analysis. The objective of this work is to directly identify the clusters, automatically generating their scatter plot and clustered trace of size no larger than a user specified maximum.

The major problem that on-line clustering introduces is the duration of the analysis when a large number of points (typically above 50000) have to be clustered. In order to obtain a faster characterization of the application we sample a subset of the points, cluster them and then perform a nearest neighbor classification of all other points. If the sample is sufficiently small and representative the process should result in a good characterization of the application with a significantly faster execution time than the full clusterization of all the points. In the paper we evaluate different approaches to obtain a representative sample. A good way to achieve the desired characteristics of such sample is to keep all the points of some randomly selected processes plus a random sample of the points of all other processes. A sample of a size around 15% of the original set of points does result in very good characterization of all the relevant computation bursts in the trace.

The second issue addressed in the paper looks at how to detect that an application has entered a stable phase. By monitoring the raw data production rate of the application, the on-line analysis estimates the duration of the interval that would result in a trace of the size specified as target by the user. Such an interval is used by the MRNET root process to drive the instrumentation package at each process to send their captured information. The sampling process can take place in the intermediate nodes in the MRNET tree. The clustering takes place in the root. By comparing two successive clusterings the decision is made whether the application has reached a stable state. If not, a new acquisition period is started. If a stable behavior is identified, the root tells the leaves to dump the trace for the last period. Classification can take place in the leaves and the trace merging process could also be done through the MRNET tree.

It is quite natural that all the techniques described in previous sections are related and complementary. This work started by looking at the on-line use of the clustering techniques but it is clear that future work will further integrate the spectral analysis techniques described in section 9.3 and the sampling based techniques of section 9.5.

Other work in the spirit of on-line analysis is the integration of the sampling based techniques described in section 9.5 in the TAU [8] profile based environment. Figure 9.9 used in section 9.5 actually contains some preliminar results of such work.

9.7 Conclusion

This paper presents some attempts to leverage ideas and techniques from different areas such as signal processing and data mining in the area of performance analysis tools. It is based on the believe that a huge body of theory and experience has been developed in other fields that has not yet been applied to enable very precise and fine grained automatic analysis of application performance.

We have described the use of spectral analysis techniques and mathematical morphology to identify how long should we trace an application to obtain full detail of its behavior. Clustering has been applied to identify structure within an application and obtain from a single run very complete and precise statistics of all hardware counter metrics. The combined use of instrumentation and sampling has been used to demonstrate that it is possible to obtain extremely precise information of the evolution of instantaneous performance metrics such as mips without incurring overheads.

We do believe that these techniques will be further improved in the future but their potential is huge. Used in combination with many other techniques this will help evolve performance tools in the direction of minimizing the amount of data emitted by on-line monitoring but providing much more information than what is today's practice.

Acknowledgements This chapter summarizes the views of the author, but is based on work by many other people in the BSC tools team (specially Judit Gimenez, Harald Servat, Juan Gonzalez, German Llort and Kevin Huck). The work is partially funded by the IBM, through the IBM-BSC Mareincognito collaboration agreement, the USAF grant FA8655-09-1-3075 and the Spanish Ministry of Education under grant TIN2007-60625.

References

1. H. Servat, G. Llort, J. Gimenez, J. Labarta: Detailed performance analysis using coarse grain sampling 2nd Workshop on Productivity and Performance. PROPER 2009.
2. J. Gonzalez, J. Gimenez and J. Labarta: Automatic Detection of Parallel Applications Computation Phases. Proceedings of the 23rd IEEE International Parallel and Distributed Processing Symposium (IPDPS'09), (2009)
3. J. Gonzalez, J. Gimenez and J. Labarta: Automatic evaluation of the computation structure of parallel applications. PDCAT 2009.
4. Casas, M.; Badia, R. M.; Labarta, J. Automatic Structure Extraction from MPI Applications Tracefiles. Euro-Par 2007. 3–12

5. J. Labarta, S. Girona, V. Pillet, T. Cortes and L. Gregoris: DiP: A Parallel Program Development Environment. Proc. of 2nd International EuroPar Conference (EuroPar 96) (1996)
6. W. E. Nagel, A. Arnold, M. Weber, H. C. Hoppe and K. Solchenbach: VAMPIR: Visualization and Analysis of MPI Resources. Supercomputer, vol. 12, n. 1, 69–80, (1996).
7. G. Llort, J. Gonzalez, H. Servat, J. Gimenez and J. Labarta. On-line detection of large-scale parallel application's structure IPDPS 2010.
8. S. Shende and A. D. Malony: The TAU Parallel Performance System. International Journal of High Performance Computing Applications, Volume 20 Number 2 Summer 2006. 287–311
9. P. C. Roth, D. C. Arnold, and B. P. Miller: MRNet: A Software-Based Multicast/Reduction Network for Scalable Tools. SC2003, Phoenix, Arizona, November 2003
10. Browne, S., Dongarra, J., Garner, N., London, K., Mucci, P.: A Scalable Cross-Platform Infrastructure for Application Performance Tuning Using Hardware Counters. Proceedings of SuperComputing 2000 (SC'00), Dallas, TX, November 2000
11. A. Mericas et al.: CPI analysis on POWER5, Part 2: Introducing the CPI breakdown model. `https://www.ibm.com/developerworks/library/pa-cpipower2/`
12. Labarta J., Gimenez J.: Performance Analysis: From Art to Science. In Parallel Processing for Scientific Computing. M. Heroux and R. Raghavan and H.D. Simon Eds. SIAM. 2006. 9–32.

Chapter 10
The Importance of Run-Time Error Detection

Glenn R. Luecke, James Coyle, James Hoekstra, Marina Kraeva, Ying Xu, Mi-Young Park, Elizabeth Kleiman, Olga Weiss, Andre Wehe and Melissa Yahya

Abstract The ability of system software to detect and issue error messages that help programmers quickly fix serial and parallel run-time errors is an important productivity criterion for developing and maintaining application programs. Over ten thousand run-time error tests and a run-time error detection (RTED) evaluation tool has been developed for the automatic evaluation of run-time error detection capabilities for serial errors and for parallel errors in MPI, OpenMP and UPC programs. Evaluation results, tests and the RTED evaluation tool are freely available at `http://rted.public.iastate.edu`. Many compilers, tools and run-time systems scored poorly on these tests. The authors make recommendations for providing better RTED in the future.

Key words: Run-time error detection, Fortran, C, CH, MPI, OpenMP, UPC

10.1 Introduction

Debugging serial and parallel programs can be very time consuming. The typical debugging process is: (1) first compile and correct all compile-time errors, (2) next run and correct the run-time errors issued by the run-time system and (3) then use a debugger and/or print statements to find and correct the rest of the errors, i.e. the errors not detected at run-time and logical errors. Compile-time errors can normally be corrected quickly since compilers usually issue good error messages. Similarly, usually run-time errors can be corrected quickly without a debugger and print statements when the run-time system correctly diagnoses the error and issues a good error message. However, correcting the other errors can be very time consuming. If a run-time system does not issue a good error message, then one is forced to use a debugger and/or print statements to find and correct the error. Notice that print

Glenn R. Luecke, James Coyle, James Hoekstra, Marina Kraeva
Iowa State University's High Performance Computing Group, Iowa State University, Ames, Iowa 50011, USA, e-mail: {grl, jjc, hoekstra, kraeva}@iastate.edu

M.S. Müller et al. (eds.), *Tools for High Performance Computing 2009*,
DOI 10.1007/978-3-642-11261-4_10, © Springer-Verlag Berlin Heidelberg 2010

statements and debuggers only give values of variables and it is up to the person debugging the program to know if these values are correct or not. Thus, high quality run-time error detection with high quality run-time error messages is critical to providing a productive computing environment. In addition, high quality run-time error detection would be especially valuable for petascale computing when conventional parallel debuggers may not scale to hundreds of thousands of cores.

The productivity enhancement from having excellent run-time error detection (RTED) depends on many factors; e.g., the type of error, the length and complexity of the program, the experience and intelligence of the person trying to debug the program as well as this person's knowledge of the program. A few years ago, an expert in parallel programming and professor of physics at Iowa State University (ISU) spent nine months trying to find what was causing the physics code that he wrote to abort after several days of execution. He then asked ISU's HPC Group to help find the error in his 6,000 line, Fortran-MPI application code. The code was run using the MPI-CHECK tool [12]. The error was detected and a good error message was issued. With this information, the professor was able to correct the error quickly. This physics code is now running on machines all over the world and is being run regularly with 30,000 processors. Without a run-time error detection tool, the error might have never been found.

With funding from the US Department of Defense, from DARPA's High Productivity Computing Initiative and from the Extreme Scale System Center at Oak Ridge National Laboratory, extensive run-time error tests have been developed from 2003 through 2008 by Iowa State University's High Performance Computing Group for evaluating run-time error detection capabilities for

- serial errors in programs written in Fortran, C and C++ (2695 tests)
- parallel MPI errors in programs written in Fortran, C and C++ (1942 tests)
- parallel OpenMP errors in programs written in Fortran, C and C++ (3307 tests)
- parallel programs written in Unified Parallel C (2247 tests).

More than ten thousand run-time error tests have been written for this project. A run-time error detection (RTED) evaluation tool has been developed for running these tests and for automatically evaluating the run-time error messages generated by assigning a score from 0 to 5 based on the usefulness of the message to help fix the error quickly. The tool then automatically averages these scores over the different error categories and reports the results. These tests and the RTED evaluation tool provide an easy way to evaluate and compare run-time error detection capabilities provided by different vendors and could be used as part of a computer procurement process. In addition, vendors could use these tests, recommended error messages and the RTED evaluation tool to evaluate and improve the run-time error detection capabilities of their compilers, tools and run-time systems.

Current test results, tests, desired output files, and the RTED evaluation tool are freely available at `http://rted.public.iastate.edu`. As new compilers, tools and run-time systems become available, vendors are encouraged to send the results using the new system software to `rted.project@iastate.edu` so they can be posted on this web site.

10.2 Background

There are many commercial and public domain software systems to detect and provide information to help programmers fix serial run-time errors. The survey in [1] found that the commercial products, *Insure++* and *Purify* performed by far the best of all software systems evaluated. At the time of the study, *Insure++* was considered better than *Purify* but both products did an excellent job in detecting serial run-time errors in C and C++ and provided excellent information for the quick fixing of the errors. Unfortunately, neither *Insure++* nor *Purify* find run-time errors for Fortran. Sun's HPC ClusterTools [2] contains the *bcheck* tool for finding serial run-time errors in Fortran, C and C++ programs.

The Message Passing Interface, MPI, is the standard message passing library used for writing parallel scientific programs for distributed memory parallel computers [3, 4]. OpenMP is often used for writing parallel scientific programs for shared memory parallel computers [5, 6]. Since most of today's parallel computers are a collection of shared memory compute nodes interconnected via a communication network, some application programs are written using both MPI and OpenMP and some using only MPI. Unified Parallel C [7, 8] is an extension of C for parallel execution on shared or distributed memory parallel machines. Some scientific applications are written entirely in UPC instead of MPI and/or OpenMP.

There are several tools to aid in the debugging of MPI programs. The Umpire tool [9, 10] was initially developed by Jeffrey Vetter and Bronis de Supinski in 2000 at Lawrence Livermore National Laboratory. The High Performance Computing Center (HLRS) in Stuttgart and the Technische Universitaet Dresden (ZIH) in Germany have developed the *MARMOT* tool [11] for finding MPI run-time errors in Fortran and C programs. Iowa State University's High Performance Computing Group has developed MPI-CHECK for Fortran [12]. Intel's *message checker* [13] is a tool that has been developed to find MPI run-time errors in Fortran, C and C++ programs. Intel has integrated *message checker* into their *trace analyzer* and *collector* 2.7 tools within their *Cluster Toolkit*.

Intel's *thread checker* [14] and Sun Microsystems' *thread analyzer* [2] are tools for debugging OpenMP run-time errors. The authors are not aware of any run-time error detection tools for Unified Parallel C (UPC).

10.3 Methodology

This section summarizes the methodology used to develop the run-time error tests and the software for the automatic running of tests and for the automatic evaluation of the error messages. For each run-time error, test programs have been written to determine if the error can be detected and a quality message generated (each test program contains one and only one run-time error). Tests were written so that the information needed to detect the error is not available at compile-time. For each test a file with a recommended error message was created that contains the error name, the line number and the file name where the error occurred along with any additional information that would assist a programmer to find and correct the error.

The following are the run-time error categories used for the development of the serial tests for Fortran, C and C++: array index out of bounds, uninitialized variables, subprogram call errors, pointer errors, floating point errors, string errors, allocation and deallocation errors, memory leaks, input and output errors, Fortran 95 specific errors, Fortran array conformance errors and C99 specific errors.

The following are the run-time error categories used for the development of the MPI tests for Fortran, C and C++: buffer out of bounds, buffer overlap, data type errors, rank errors, other argument errors, wrong order of MPI calls, negative message length, deadlocks, race conditions, implementation dependent errors (potential deadlocks, race conditions and noncontiguous dynamic allocation of message buffers in C).

The run-time error categories used for the Fortran, C and C++ OpenMP tests are: deadlocks, race conditions, environment and clause errors, wrong order of OpenMP directives, uninitialized shared and private variables, wrong usage of OpenMP run-time library routines and implementation dependent errors (i.e. behavior is either undefined or said to be implementation dependent by the OpenMP API).

The run-time error categories used for UPC are: out-of-bounds shared memory access using indices, out-of-bounds shared memory access using pointers, out-of-bounds shared memory access in UPC functions, argument errors in UPC functions, wrong order of UPC calls, uninitialized variables, deadlocks, race conditions, memory related errors, undefined UPC operations, warnings. The "warnings" category includes tests where programmers should be warned of likely errors. For example, it makes no sense to have the "nelems" argument in reduction functions be zero even though the UPC specification allows this, so some tests have nelems = 0 and the RTED evaluation tool checks whether good warning messages are produced.

The RTED evaluation tool mentioned in Section 1 is a collection of scripts for the automatic running of tests, comparing actual messages with expected messages and then assigning a score of 0, 1, 2, 3, 4 or 5 to the message generated for each test:

- A score of 5 is given for a detailed error message that allows for the quick fixing of the error. For example, when there is an out-of-bounds access of the second dimension of array B, instead of issuing a message "out-of-bounds access in array B in line 1735 of file prog.f90", the message could mention that the problem occurred in the second dimension of B, the accessed value was 17 and the allocated range was from 1 to 16 and that B was allocated in line 923 of prog.f90.
- A score of 4 is given for error messages with more information than a score of 3 and less than 5. This is tailored for each test.
- A score of 3 is given for error messages with the correct error name, line number and the name of the file where the error occurred.
- A score of 2 is given for error messages with the correct error name and line number where error occurred but not the file name where the error occurred.
- A score of 1 is given for error messages with the correct error name.
- A score of 0 is given when the error was not detected.

Different compilers, tools and run-time systems may issue different messages (with different error names) for the same run-time error. For example, the out-of-bounds access is named as "array out-of-bounds error" on one system, "array index error" on another system and "index out-of-bounds error" on a third system. The RTED evaluation tool has a list of synonymous phrases for each error so that equivalent error messages will be evaluated appropriately. Thus, for the RTED tool to accurately evaluate an error message, the error name must be listed as one of the RTED synonymous phrases. Error messages are evaluated as follows:

- For each test and score a scoring script was created.
- A synonym file of acceptable error names was created.
- Error messages are reduced to a canonical form for easy comparison with the recommended error messages by first changing all text to lower case and then replacing selected phrases with standard phrases. Blanks, hex addresses, and integers longer than three digits are removed to reduce false matches.
- Scoring scripts are applied to the canonical form of each error message for automatic evaluation.

10.4 Results

This section contains the results of running all the tests using the software environments/machines that were available to us. Tests were run using all available compiler run-time error detection options. For each compiler, we searched the man pages and selected all debugging options for this evaluation. It would be helpful if compilers had a general "–debug" option that would turn on all debugging options. Results on the web site, `http://rted.public.iastate.edu`, present scores for each category of run-time errors. Due to space limitations, we cannot present all of the RTED results and only present average scores in most cases. Some vendors score well in some error categories and poorly in others. When taking averages, this information may be hidden, so the reader is encouraged to view the complete results posted on the web site.

There are 1552 serial execution Fortran tests, 716 serial execution C tests and 1143 serial execution C++ tests. (We ran the 716 C tests and 427 C++ specific tests for our 1143 C++ tests.) Table 10.1 presents the average scores when running these tests on different machines/software environments.

Notice that for the serial C and C++ tests, Insure++ is the only one that scored well. The Cray X1 and the NAG Fortran compilers both scored well.

There are 744 MPI Fortran tests, 723 MPI C tests and 1198 MPI C++ tests. Table 10.2 presents the average scores when running the MPI 2.0 tests on different machines/software environments. The RTED web site contains the scores for both MPI 1.1 and for MPI 2.0, but only the MPI 2.0 results are presented in this paper. If an MPI implementation does not support the full MPI 2.0 standard, there is no penalty when running the MPI 2.0 tests. This is because the RTED evaluation tool only assigns a score to those tests that compile and link successfully and average

Table 10.1: Average serial execution results.

Compiler/tool	Fortran	C	C++
Cray XT4 CNL, Pathscale compilers	0.62	0.08	0.06
Cray X2 CNL, Cray compilers	2.07	0.39	0.47
Cray X1 Unicos/mp, Cray compilers	2.49	0.47	0.53
Cray XT4 CNL, PGI compilers	1.37	0.43	0.28
IBM AIX, XLF/XLC compilers	1.51	0.12	0.10
RedHat Linux, NAGWare Fortran 95	2.36	NA	NA
RedHat Linux, Intel compilers	1.34	0.00	0.00
RedHat Linux, Intel compilers with Insure++	NA	2.75	2.97
SUN Solaris, Sun compilers with bcheck	2.11	1.08	1.29
SUN Solaris, Sun compilers	1.79	0.00	0.00
GNU v4.1.2 compilers	1.19	0.08	0.06

scores are calculated on the reduced set of tests. Notice that *MPI-CHECK* and *Marmot* scored better than the others. Intel's *trace analyzer* was not able to identify the line number where the error occurred so the best possible score for each test would be 1.0.

Table 10.2: Average MPI 2.0 results.

Compiler/tool	Fortran	C	C++
Cray X1 Unicos/mp, Cray compilers	0.64	0.75	0.82
Cray XT4 CNL, MPICH2, PGI compilers	0.24	0.32	0.31
IBM AIX, XLF/XLC compilers	0.40	0.46	0.43
RedHat Linux, OpenMPI, Pathscale	0.25	0.32	0.30
RedHat Linux, OpenMPI, Pathscale with MPI-CHECK	1.32	NA	NA
RedHat Linux, OpenMPI, Pathscale with Marmot	1.27	1.35	0.83
RedHat Linux, MPICH-gm, Intel compilers	0.25	0.46	0.43
Suse, MPICH 1.2, Intel compilers with Trace Analyzer	0.47	0.50	0.48
SUN Solaris, MPICH2, Sun compilers	0.29	0.34	0.32

Since MPI is the most commonly used method of parallelization, table 10.3 presents the detailed MPI 2.0 results for Fortran and for selected compilers, MPI Libraries and tools. Cray XT results are for the Cray XT4 system, using MPICH2 library and PGI Fortran compiler. Intel TA results were obtained on a Xeon cluster running MPICH 1.2 using Intel compiler along with the Intel Trace Analyzer and Collector tool. OpenMPI results are for an Opteron cluster, using OpenMPI library and Pathscale Fortran compiler. The results in the last two columns were obtained on the same Opteron cluster, using the MPI-CHECK and Marmot tools.

Table 10.3: MPI 2.0 Fortran results for each error category.

MPI Error Category	Cray XT	Intel TA	OpenMPI	MPI-CHECK	Marmot
Buffer out of bounds	0.01	0.06	0.02	2.13	0.05
Buffer overlap	0.00	0.71	0.05	0.62	0.14
Datatype errors	0.10	0.40	0.20	1.80	0.20
Rank errors	0.54	0.76	0.08	2.81	1.87
Other argument errors	0.22	0.59	0.00	0.30	0.37
Wrong order of MPI calls	0.48	0.69	0.06	0.56	1.48
Negative message length	0.98	0.91	0.08	1.95	1.42
Deadlocks	0.02	0.17	0.00	1.79	2.69
Race conditions	0.00	0.06	0.00	0.09	1.07
Implementation dependent errors	0.04	0.31	0.00	3.18	3.39
AVERAGES	0.24	0.47	0.05	1.52	1.27

The following is one of the Fortran MPI tests that we wrote:

```
program F_C_1_3_1_2_e_M1
   implicit none
   include "mpif.h"
   integer, parameter :: N=5      ! buffer size
   integer :: arrayA(N), arrayB(N+1), sbuf(N+1)
   call MPI_INIT(ierror)
   call MPI_COMM_SIZE(mpi_comm_world, numprocs,
                            ierror)
   call MPI_COMM_RANK(mpi_comm_world, myrank,
                            ierror)
   if(cos(x) > 2.0) then
      count = N
   else
      count = N+1
   endif

   do i=1,count
      sbuf(i) = myrank + i
   enddo

   if(myrank.eq.1) then
      call MPI_ALLREDUCE(sbuf, arrayA, count,
      MPI_INTEGER, MPI_SUM, mpi_comm_world, ierror)
   else
      call MPI_ALLREDUCE(sbuf, arrayB, count,
      MPI_INTEGER, MPI_SUM, mpi_comm_world, ierror)
   endif
   call MPI_FINALIZE(ierror)
end program F_C_1_3_1_2_e_M1
```

This program reads the value of x from a file so that its value is not known to the program at compile time. For this example Pathscale with OpenMPI, Pathscale with OpenMPI and Marmot, and Intel's Trace Analyzer and Collector ail did not detect the error and were given a score of zero. MPI-CHECK with the Pathscale compiler and OpenMPI received a score of 4.0 for producing the following message:

> File=/scratch/jjc/F_C_1_3_1_2_e_M1.f90, Line= 76, Argument= 2] arrayA, message size exceeds the bounds of this array, please check the message size.

The recommended error message for this test is:

> Buffer size exceeded. The value 6 of argument 'count' in 'MPI_ALLREDUCE' called at line 76 in file 'F_C_1_3_1_2_e_M1.f90' on process 1 exceeds the size of receive buffer 'arrayA'. 'arrayA' is declared at line 48 in file 'F_C_1_3_1_2_e_M1.f90' with size 5.

There are 2156 OpenMP Fortran tests, 1066 OpenMP C tests, and 1 151 OpenMP C++ tests. (We ran the 1066 C tests and 85 C++ specific tests for our 1151 C++ tests.) There are few C++ specific tests since the OpenMP API has few items that are C++ specific. Table 10.4 presents the average scores when running these tests on different machines/software environments. Notice that Intel's *thread checker* and Sun's *thread analyzer* both improved the score, but not by much.

Table 10.4: Average OpenMP results.

Compiler/tool	Fortran	C	C++
Cray X1 Unicos(mp), Cray compilers	0.32	0.30	0.45
Cray X2 Unicos(mp), Cray compilers	0.23	0.25	0.40
Cray Unicos, PGI compilers	0.17	0.19	0.21
Cray Unicos, GNU compilers	0.20	0.19	0.27
Cray Unicos, Pathscale compilers	0.13	0.18	0.21
IBM AIX, XLF/XLC compilers	0.26	0.23	0.30
RedHat Linux, Pathscale compilers	0.13	0.19	0.24
RedHat Linux, Intel compilers	0.13	0.14	0.20
RedHat Linux, Intel compilers with thread checker	0.42	0.43	0.52
SUN Solaris, Sun compilers	0.02	0.02	0.03
SUN Solaris, Sun compilers with thread analyzer	0.40	0.39	0.40

Table 10.5 presents the scores when running the 2247 UPC tests using Cray's, Berkeley's, HP's and GNU's UPC compilers. The section "Undefined UPC Operations" contains all situations where the outcome of certain UPC statements is stated as being undefined by the UPC specification. The UPC "warnings" category is described in section 10.3. GNU's and HP's UPC do not support UPC IO so these tests were skipped when using these compilers and scores calculated on the reduced set of tests. In addition, GNU's UPC does not support UPC collective utilities so these tests were also skipped when using GNU's UPC compiler and scores calculated on the reduced set of tests. Notice that Cray's UPC compiler scored better than all the others in some categories.

Table 10.5: UPC results for each error category.

UPC Error Category	Cray	Berkeley	HP	GNU
Out-of-bounds shared memory access using indices	1.30	0.00	0.03	0.20
Out-of-bounds shared memory access using pointers	1.04	0.00	0.00	0.21
Out-of-bounds shared memory access in UPC calls	0.91	0.00	0.02	0.01
Argument errors in UPC functions	0.38	0.04	0.00	0.00
Wrong order of UPC calls	0.84	0.20	0.53	0.89
Uninitialized variables	0.08	0.02	0.57	0.25
Deadlocks	0.00	0.58	0.36	0.27
Race conditions	0.01	0.00	0.00	0.00
Memory related errors	0.18	0.00	0.16	0.37
Undefined UPC operations	0.19	0.21	0.15	0.41
Warnings (uninitialized shared variables)	0.27	0.00	0.00	0.00
AVERAGES	0.47	0.10	0.17	0.24

10.5 Recommendations

The ability to detect and issue high quality run-time error messages is critical for programmer productivity and should be an integral part of providing a productive environment for the development and maintenance of scientific applications. In this section, we make recommendations on how this could be accomplished.

Ideally, each vendor should provide high-quality RTED. However, the results of this study show that this is not the current situation and there are no signs that this will change. Since JAVA's language specification includes array bounds checking, we thought that RTED could be part of the Fortran, C and C++ language specifications as a first step towards providing high quality RTED. Vendors could then use the RTED tests developed in the project when implementing RTED. This idea was presented to the Fortran standards committee. After discussion the idea was rejected. The idea has not been presented to the C and C++ standards committees.

Since high quality RTED is so important for the productivity of application code developers, we now recommend the development of high quality RTED tools that are freely available and support each commonly-used programming paradigm. Funding for these tools must include not only their development but also ongoing maintenance, periodic enhancement with better RTED techniques and with programming paradigm advancements. The following lists the specific programming paradigms we recommend RTED tools be developed for:

- Serial Fortran, C, C++
- MPI with Fortran, C and C++
- OpenMP with Fortran, C and C++
- UPC
- Co-Array Fortran

Since MPI programs are Fortran, C or C++ programs calling MPI functions, the MPI tool should be used along with the serial tools for Fortran, C and C++. The OpenMP tools could be developed from the serial Fortran, C and C++ tools.

Since NVIDIA is providing CUDA for programming for their GPU, we recommend NVIDIA fund the development and maintenance of RTED tools for CUDA.

There are three new parallel languages that have been developed as part of DARPA's High Productivity Computing Initiative: Chapel developed by Cray, Fortress developed by Sun Microsystems and Q10 developed by IBM. We recommend that each of these vendors fund the development and maintenance of a high quality RTED tool for their own language.

10.6 Conclusions

The ability of system software to detect errors at run-time and issue error messages that help programmers quickly fix errors is an important productivity criterion for developing and maintaining application programs. Over ten thousand run-time error tests and a run-time error detection (RTED) evaluation tool have been developed for the automatic evaluation of run-time error detection capabilities for serial errors and for parallel errors in MPI, OpenMP and UPC programs. Each error message issued by the run-time system is assigned a score from 0 to 5 based on the usefulness of the information in the message to help a programmer quickly fix the error. Average scores over error categories are automatically calculated and reported. All tests and the RTED evaluation tool are freely available at the RTED web site `http://rted.public.iastate.edu`. Many compilers, tools and run-time systems have been evaluated with results posted on this same web site.

The technology for detecting and reporting many run-time errors is known, but the results of running these tests show that many of the software environments evaluated currently do a poor job detecting run-time errors with the following exceptions:

- For the serial tests, Insure++ scored well for C and C++ programs and the Cray X1 and NAG compilers both scored well for Fortran.
- For the MPI tests, MPI-CHECK and Marmot scored better than the others.

It is hoped that these tests and recommended error messages will be used to evaluate and improve the run-time error detection capabilities of compilers, tools and run-time systems and that these tests will also be used by high performance computing centers as part of their computer procurement process.

The authors recommend the development of high-quality, public domain RTED tools to support the programming paradigms commonly used for scientific computing. Funding for these projects should include not only development but also maintenance, periodic enhancements with better RTED techniques and support future programming paradigm enhancements.

Acknowledgements This work was funded by the US Department of Defense, DARPA's High Productivity Computing Initiative and by the Extreme Scale System Center at Oak Ridge National Laboratory.

References

1. Luecke, G., Coyle, J., Hoekstra, J., Kraeva, M., Li, Y., Taborskaia, O., Wang, Y.: A Survey of Systems for Detecting Serial Run-time Errors. Concurrency and Computation: Practice and Experience, vol. 18, pp 1885–1907 (2006)
2. Sun Microsystem's HPC ClusterTools, http://www.sun.com/software/products/clustertools/
3. Snir, M., Otto, S. W., Huss-Lederman, S., Walker, D. W., Dongarra, J.: MPI - The Complete Reference, The MIT Press (1998)
4. Message Passing Interface Forum, http://www.mpi-forum.org
5. The OpenMP API Specification, http://openmp.org
6. Chapman, B., Jost, G., Van der Pas, R.: Using OpenMP: Portable Shared Memory Parallel Programming, The MIT Press (2008)
7. Unified Parallel C, http://upc.gwu.edu
8. El-Ghazawi, T., Carlson, W., Sterling, T., Yelick, K.: UPC Distributed Shared Memory Programming, Wiley-Interscience (2005)
9. Vetter, J.S., De Supinski, B.R.: Dynamic software testing of MPI applications with Umpire, In: Conference on High Performance Networking and Computing Article 51, Proceedings of the 2000 ACM/IEEE conference on Supercomputing, Dallas, Texas, United States (2000)
10. Hilbrich, T., Supinski, B., Mueller, M., Schulz, M.: A Graph Based Approach for MPI Deadlock Detection, In: International Conference on Supercomputing, Yorktown Heights, NY, USA, pp 296–305 (2009)
11. MARMOT, http://www.hlrs.de/organization/av/amt/research/marmot/publications/
12. Luecke, G.R., Chen, H., Coyle, J., Hoekstra, J., Kraeva, Zou, Y.: MPI-CHECK: a Tool for Checking Fortran 90 MPI Programs. Concurrency and Computation: Practice and Experience, vol. 15, pp 93–100 (2003)
13. Intel Message Checker, http://www.intel.com/cd/software/products/asmo-na/eng/227074.htm
14. Intel Thread Checker, http://software.intel.com/en-us/intel-thread-checker/

Chapter 11
Collecting Performance Data with PAPI-C

Dan Terpstra, Heike Jagode, Haihang You, and Jack Dongarra

Abstract Modern high performance computer systems continue to increase in size and complexity. Tools to measure application performance in these increasingly complex environments must also increase the richness of their measurements to provide insights into the increasingly intricate ways in which software and hardware interact. PAPI (the Performance API) has provided consistent platform and operating system independent access to CPU hardware performance counters for nearly a decade. Recent trends toward massively parallel multi-core systems with often heterogeneous architectures present new challenges for the measurement of hardware performance information, which is now available not only on the CPU core itself, but scattered across the chip and system. We discuss the evolution of PAPI into Component PAPI, or PAPI-C, in which multiple sources of performance data can be measured simultaneously via a common software interface. Several examples of components and component data measurements are discussed. We explore the challenges to hardware performance measurement in existing multi-core architectures. We conclude with an exploration of future directions for the PAPI interface.

11.1 Introduction

The use of hardware counters to measure and improve software performance has become an accepted and integral method in the software development cycle [1]. Hardware counters, which are usually implemented as a small set of registers onto

Dan Terpstra, Heike Jagode, Jack Dongarra
The University of Tennessee, e-mail: {terpstra, jagode, dongarra}@eecs.utk.edu

Jack Dongarra
Oak Ridge National Laboratory

Haihang You
National Institute for Computational Sciences
e-mail: you@eecs.utk.edu

M.S. Müller et al. (eds.), *Tools for High Performance Computing 2009*,
DOI 10.1007/978-3-642-11261-4_11, © Springer-Verlag Berlin Heidelberg 2010

which can be mapped a larger set of performance related events, can provide accurate and detailed information on a wide range of hardware performance metrics. PAPI, the Performance Application Programming Interface, provides an easy to use, common API to application and tool developers to supply them with the information they may need to analyze, model and tune their software on a wide range of different platforms.

In addition to the counters found on CPUs, a large amount of hardware monitoring information is also available in other sub-sytems throughout modern computer architectures. Many network switches and network interface cards (NICs) contain counters that can monitor various events related to performance and reliability. Possible events include checksum errors, dropped packets, and packets sent and received. Although the set of network events is necessarily somewhat dependent on the underlying hardware, extending PAPI to the network monitoring domain can provide a portable way to access native network events and allow correlation of network events with other domains. Because communication in OS-bypass networks such as Myrinet and Infiniband is handled asynchronously to the application, hardware monitoring, in addition to being low overhead, may be the only way to obtain some important data about communication performance.

As processor densities climb, the thermal properties and energy usage of high performance systems are becoming increasingly important. Such systems contain large numbers of densely packed processors which require a great deal of electricity. Power and thermal management issues are becoming critical to successful resource utilization [2, 3]. Standardized interfaces for accessing the thermal sensors are available, but may be difficult to use for runtime power-performance adaptation [4]. Extending the PAPI interface to simultaneously monitor processor metrics and thermal sensors can provide clues for correlating algorithmic activity with thermal system responses thus help in developing appropriate workload distribution strategies. We show the results of using the extended version of PAPI to simultaneously monitor processor counters, ACPI thermal sensors, and Myrinet network counters while running the FFTE and HPL HPC Challenge benchmarks [12] on an AMD Opteron Linux cluster.

Modifying and extending a library with a broad user base such as PAPI requires care to preserve simplicity and backward compatibility as much as possible while providing clean and intuitive access to important new capabilities. We discuss modifications to PAPI to provide support for the simultaneous measurement of data from multiple counter domains.

With the advent of multi-core processors and the inexorable increase in core counts per chip, interactions between cores and contention for shared resources such as last level caches or memory bus bandwidth become increasingly important sources of potential performance bottlenecks. Individual vendors have chosen different paths to provide access to hardware performance monitoring for these shared resources, each with their own problems and issues. We explore some of these approaches and their implications for performance measurement, and provide an example measurement of cache data on a real application in a multi-core environment to illustrate these issues.

11.2 Extending PAPI to Multiple Measurement Components

The PAPI library was originally developed to address the problem of accessing the processor hardware counters found on a diverse collection of modern microprocessors in a portable manner [1]. Other system components besides the processor, such as heterogeneous processors (GPUs), memory interface chips, network interface cards, and network switches, also have hardware that counts various events related to system reliability and performance. Furthermore, other system health measurements, such as chip or board level temperature sensors, are available and useful to monitor in a portable manner. Unlike on-processor counters, the off-processor counters and sensors usually measure events in a system-wide rather than a process or thread-specific context. However, when an application has exclusive use of a machine partition, or runs in a single core of a multi-core node, it may be possible to interpret such events in the context of the application. Even with execution on multiple cores on a single node it may be possible to deconvolve the temperature or power signatures of separate threads to develop a coarse picture of single thread response. The current situation with off-processor counters is similar to the situation that existed with on-processor counters before PAPI. A number of different platform-specific interfaces exist, some of which are poorly documented or not documented at all.

Several software design issues became apparent in extending the PAPI interface for multiple measurement domains. The classic PAPI library consists of two internal layers: a large portable layer optimized for platform independence; and a smaller hardware specific layer, containing platform dependent code. By compiling and statically linking the independent layer with the hardware specific layer, an instance of the PAPI library could be produced for a specific operating system and hardware architecture. At compile time the hardware specific layer provided common data structure sizes and definitions to the independent layer, and at link time it satisfied unresolved function references across the layers. Since there was a one-to-one relationship between the independent layer and the hardware specific layer, initialization and shutdown logic was straightforward, and control and query routines could be directly implemented. In migrating to a multi-component model, this one-to-one relationship was replaced with a one-to-many coupling between the independent, or framework, layer and a collection of hardware specific components, requiring that previous code dependencies and assumptions be carefully identified and modified as necessary.

When linking multiple components into a common object library, each component exposes a subset of the same functionality to the framework layer. To avoid name-space collisions in the linker, the entry points of each component are modified to hide the function names, either by giving them names unique to the component, or by declaring them as static inside the component code. Each component contains an instance of a structure, or vector, with all the necessary information about opaque structure sizes, component specific initializations and function pointers for each of the functions that had been previously statically linked across the framework/component boundary. The only symbol that a component exposes to the framework at link time is this uniquely named component vector. All accesses to the component code occur through function pointers in this vector, and empty vector pointers fail gracefully, allowing components to be implemented with only a subset of the complete functionality. In this way, the framework can transparently manage initialization of and access to multiple components by iterating across a list of all available component structures. Our experiments have shown that the extra level of indirection introduced by calls through a function pointer adds a small but generally negligible additional overhead to the call time, even in time-critical routines such as reading counter values. Timing tests were done on hardware including Intel Pentium4, Core2, and Nehalem, AMD Opteron and IBM POWER6 architectures. Over 1M iterations of a loop including 10 calls to empty subroutines the average execution time difference between direct and indirect calls was in the range of 6.9% for Nehalem to 46% for POWER6. In the context of real PAPI workloads on these same machines, a start/stop operation was slowed by between 0.13% and 1.36%, while a read of two counters was slowed by between 1.26% and 11.3%. Table 11.1 shows these results in greater detail.

Table 11.1: Costs of PAPI calls

	Pentium4	Core2	Nehalem	Opteron	POWER6
direct cycles/call	13.8	8.4	5.8	9.6	106.3
indirect cycles/call	17.8	10.3	6.2	11	155.2
% slowdown	29.00%	22.60%	6.90%	14.60%	46.00%
PAPI start/stop slowdown	0.66%	0.52%	0.13%	0.39%	1.36%
PAPI read 2 counters slowdown	9.76%	6.40%	2.47%	11.30%	1.26%

Countable events in PAPI are either preset events, defined uniformly across all architectures, or native events, unique to a specific component. To date preset events have only been defined for processor hardware counters, making all events on off-processor components native events.

11.2.1 Preset Events

Preset events can be defined as a single event native to a given CPU, or can be derived as a linear combination of native events, such as the sum or difference of two such events. More complex derived combinations of events can be expressed in reverse polish notation and computed at run-time by PAPI. The number of unique terms in these expressions is limited by the number of counters in the hardware. For many platforms the preset event definitions are provided in a comma separated values file, `papi_events.csv`, which can be modified by developers to explore novel or alternate definitions of preset events. Because not all preset events are implemented on all platforms, a utility called `papi_avail` is provided to examine the list of preset events on the platform of interest. A portion of the output for an Intel Nehalem (core i7) processor is shown below:

```
Available events and hardware information.
--------------------------------------------------------------------------------
PAPI Version             : 4.0.0.0
Vendor string and code   : GenuineIntel (1)
Model string and code    : Intel Core i7 (21)
CPU Revision             : 5.000000
CPUID Info               : Family: 6  Model: 26  Stepping: 5
CPU Megahertz            : 2926.000000
CPU Clock Megahertz      : 2926
Hdw Threads per core     : 1
Cores per Socket         : 4
NUMA Nodes               : 2
CPU's per Node           : 4
Total CPU's              : 8
Number Hardware Counters : 7
Max Multiplex Counters   : 32
--------------------------------------------------------------------------------
The following correspond to fields in the PAPI_event_info_t structure.

    Name        Code    Avail Deriv Description (Note)
PAPI_L1_DCM  0x80000000  No    No   Level 1 data cache misses
PAPI_L1_ICM  0x80000001  Yes   No   Level 1 instruction cache misses
PAPI_L2_DCM  0x80000002  Yes   Yes  Level 2 data cache misses
...
PAPI_FP_OPS  0x80000066  Yes   Yes  Floating point operations
PAPI_SP_OPS  0x80000067  Yes   Yes  Floating point operations;
                                    optimized to count scaled single precision
                                    vector operations
PAPI_DP_OPS  0x80000068  Yes   Yes  Floating point operations;
                                    optimized to count scaled double precision
                                    vector operations
PAPI_VEC_SP  0x80000069  Yes   No   Single precision vector/SIMD instructions
PAPI_VEC_DP  0x8000006a  Yes   No   Double precision vector/SIMD instructions
--------------------------------------------------------------------------
Of 107 possible events, 34 are available, of which 8 are derived.
```

11.2.2 Native Events

PAPI components contains tables of native event information allowing native events to be programmed in essentially the same way as a preset event. Each native event may have a number of attributes, called unit masks, that can act as filters on exactly what gets counted. These attributes can be appended to a native event name to tell PAPI exactly what to count. An example of a native event name with unit masks from the Intel Nehalem architecture is shown below:

```
L2_DATA_RQSTS:DEMAND_M_STATE:DEMAND_I_STATE
```

Attributes can be appended in any order and combination, and are separated by colon characters. Some components such as LM-SENSORS may have hierarchically defined native events. An example of such a hierarchy is shown below:

```
LM_SENSORS.max1617-i2c-0-18.temp2.temp2_input
```

In this case, levels of the hierarchy are separated by period characters. Complete listings of these and other native events can be obtained from a utility analogous to `papi_avail`, called `papi_native_avail`. A portion of the output of `papi_native_avail` for Nehalem configured with multiple components is shown below:

```
...
--------------------------------------------------------------------------------
0x40000032   L1I_OPPORTUNISTIC_HITS   | Opportunistic hits in streaming          |
--------------------------------------------------------------------------------
0x40000033   L2_DATA_RQSTS   | All L2 data requests                               |
  40000433   :ANY   | All L2 data requests                                        |
  40000833   :DEMAND_E_STATE   | L2 data demand loads in E state                   |
  40001033   :DEMAND_I_STATE   | L2 data demand loads in I state (misses)          |
  40002033   :DEMAND_M_STATE   | L2 data demand loads in M state                   |
  40004033   :DEMAND_MESI   | L2 data demand requests                              |
  40008033   :DEMAND_S_STATE   | L2 data demand loads in S state                   |
  40010033   :PREFETCH_E_STATE   | L2 data prefetches in E state                   |
  40020033   :PREFETCH_I_STATE   | L2 data prefetches in the I state (misses)      |
  40040033   :PREFETCH_M_STATE   | L2 data prefetches in M state                   |
  40080033   :PREFETCH_MESI   | All L2 data prefetches                             |
  40100033   :PREFETCH_S_STATE   | L2 data prefetches in the S state               |
--------------------------------------------------------------------------------
0x40000034   L2_HW_PREFETCH   | Count L2 HW Prefetcher Activity                  |
  40000434   :HIT   | Count L2 HW prefetcher detector hits                       |
  40000834   :ALLOC   | Count L2 HW prefetcher allocations                       |
  40001034   :DATA_TRIGGER   | Count L2 HW data prefetcher triggered             |
  40002034   :CODE_TRIGGER   | Count L2 HW code prefetcher triggered             |
  40004034   :DCA_TRIGGER   | Count L2 HW DCA prefetcher triggered               |
  40008034   :KICK_START   | Count L2 HW prefetcher kick started                 |
--------------------------------------------------------------------------------
...
--------------------------------------------------------------------------------
0x44000000   ACPI_STAT   | kernel statistics                                   |
--------------------------------------------------------------------------------
0x44000001   ACPI_TEMP   | ACPI temperature                                    |
--------------------------------------------------------------------------------
0x48000000   LO_RX_PACKETS   | LO_RX_PACKETS                                   |
--------------------------------------------------------------------------------
0x48000001   LO_RX_ERRORS   | LO_RX_ERRORS                                     |
--------------------------------------------------------------------------------
...
--------------------------------------------------------------------------------
0x4c0000b3   LM_SENSORS.w83627hf-isa-0290.cpu0_vid.cpu0_vid                      |
--------------------------------------------------------------------------------
0x4c0000b4   LM_SENSORS.w83627hf-isa-0290.beep_enable.beep_enable                |
--------------------------------------------------------------------------------
--------------------------------------------------------------------------------
Total events reported: 396
```

11.2.3 API Changes

An important consideration in extending a widely accepted interface such as PAPI is to make extensions in such a way as to preserve the original interface as much as possible for the sake of backward compatibility. Several entry points in the PAPI user API were augmented to support multiple components, and several new entry points were added to support new functionality.

By convention, an event to be counted is added to a collection of events in an *EventSet*, and *EventSets* are started, stopped, and read to produce event count values. Each *EventSet* in Component PAPI is bound to a specific component and can only contain events associated with that component. Multiple *EventSets* can be active simultaneously, as long as only one *EventSet* per component is invoked. The binding of *EventSet* and component can be done explicitly at the time it is created with a call to the new API:

```
PAPI_assign_eventset_component() - assign a component index to an
                                   existing but empty EventSet
```

Explicit binding allows a variety of attributes to be modified in an *EventSet* even before events are added to it. To preserve backward compatibility for legacy applications, binding to a specific component can also happen automatically when the first event is added to an *EventSet*.

Three entry points in the API allow access to settings within PAPI. These entry points are shown below:

```
PAPI_num_hwctrs() - return the number of hardware counters for
                    the cpu
PAPI_get_opt()    - query the option settings of the PAPI
                    library or a specific event set
PAPI_set_domain() - set the default execution domain for new
                    event sets
```

Component specific versions of these calls are:

```
PAPI_num_cmp_hwctrs() - return the number of hardware counters
                        for the cpu
PAPI_get_cmp_opt()    - query the option settings of the PAPI
                        library or a specific event set
PAPI_set_cmp_domain() - set the default execution domain for
                        new event sets
```

These modified calls have been implemented with an additional parameter to allow specification of a given component within the call. Backward compatibility is preserved by assuming that the original calls are always bound to the original cpu component.

Finally two new calls were added to provide housekeeping functions. The first simply reports the current number of components, and the second returns a structure of information describing the component:

```
PAPI_num_components()
PAPI_get_component_info()
```

Neither of these calls are required. In this way legacy code instrumented with PAPI calls compiles and runs with no modification needed.

Example components have been implemented in the initial release of PAPI for ACPI temperature sensors, the Myrinet network counters, and the lm-sensors interface. An implementation of an Infiniband network component is under investigation, along with several other components for disk sub-systems such as Lustre.

11.2.4 The CPU Component

The CPU component is unique for several reasons. Historically it was the only component that existed in earlier versions of PAPI. Within Component PAPI one and only one CPU component must exist and occupy the first position in the array of components. This simplifies default behavior for legacy applications. In addition to providing access to the hardware counters on the main processor in the system, the CPU component also provides the operating system specific interface for things like interrupts and threading support, as well as high resolution time bases used by the PAPI Framework layer. The necessity for a unique CPU component has been identified as a restriction from the perspective of implementations that may not need or wish to monitor the CPU and also implementations that may contain heterogeneous CPUs. This is an open research issue in Component PAPI and mechanisms are under investigation to relax these restrictions.

11.2.5 Accessing the CPU Hardware Counters

CPU Hardware counter access is provided in a variety of ways on different systems. When PAPI was first released almost 10 years ago, there was significant diversity in the operating systems and hardware of the Top500 list. AIX, Solaris, UNICOS and IRIX shared the list with a number of variants of Unix [8]. Linux systems made up a mere 3.6 percent of the list. Most of these systems had vendor provided support for counter access either built-in to the operating system, or available as a loadable driver. The exception was Linux, which had no support for hardware counter access. This is in sharp contrast to today [9], when nearly 90 percent of the systems run Linux or Linux variants.

Several options were available to access counters on Linux systems. One of the earliest was the `perfctr` patch [10] for x86 processors. `Perfctr` provided a low latency memory-mapped interface to virtualized 64-bit counters on a per process or per thread basis, ideal for PAPI's "first person" counting and sampling interface. With the introduction of Linux on the Itanium processor, the `perfmon` [5] interface was built-in to the kernel. When it became apparent that `perfctr` would not be accepted into the Linux kernel, `perfmon` was rewritten and generalized as `perfmon2` [11] to support a wide range of processors under Linux, including the IBM POWER series in addition to x86 and IA64 architectures. After a continuing effort over several years by the performance community to get `perfmon2` accepted into the Linux kernel, it too was rejected and supplanted by yet another abstraction of the hardware counters, first called `perf_counters` in kernel 2.6.31 and then `perf_events` [6] in kernel 2.6.32. The `perf_events` interface is young and maturing rapidly. It has the overwhelming advantage of being built-in to the kernel, requiring no patching on the part of system administrators. PAPI continues to support hardware counter access through `perfctr` wherever it is available. `Perfmon` access is available through the 2.6.30 kernel. In addition, PAPI also supports the `perf_events` interface.

11.2.6 The ACPI and MX Components

The ACPI component enables the PAPI-C library to access the ACPI temperature sensors, while the MX component allows monitoring of run-time characteristics of the Myrinet network communications. To demonstrate simultaneous monitoring of CPU metrics as well as temperature and data transfer, we collected data from the HPC Challenge suite. This suite is a set of scalable, computationally intensive benchmarks with different memory access patterns that examine the performance of HPC architectures [12]. For our experiments, we chose two global kernel benchmarks, High Performance Linpack (HPL) and FFT. The HPL kernel solves a linear system of equations and the FFT kernel computes a double precision complex one-dimensional discrete Fourier transform, which ensures two highly computationally intense test cases. We instrumented both benchmarks to gather total floating-point operations, temperature and packets sent and received through the Myrinet network. With Component PAPI, we were able to easily instrument the program by simply providing the desired event names in PAPI calls. We ran our experiments on a 65-node AMD Opteron cluster. Both benchmarks ran on eight nodes. We instrumented functions fft235 in FFT and pdgesvK2 in HPL, since profiling indicated that these were the most computationally active routines, and gathered data for each iteration that called these functions.

The measurements for the FFT benchmark on two of the nodes are shown in Fig. 11.1. We can see the periodic nature of the computation and communication. The measured data for the second case study - the HPL benchmark - is depicted in Fig. 11.2 and shows a completely different computation and communication pattern. In both test cases, we are able to observe a difference in the temperature between the two nodes.

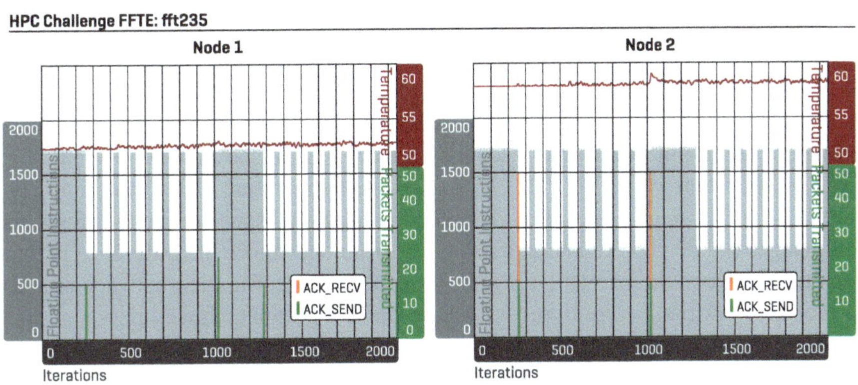

Fig. 11.1: FLOPS, temperature and communication monitoring using the CPU, ACPI and MX component of PAPI-C for an FFT benchmark running on an AMD Opteron cluster

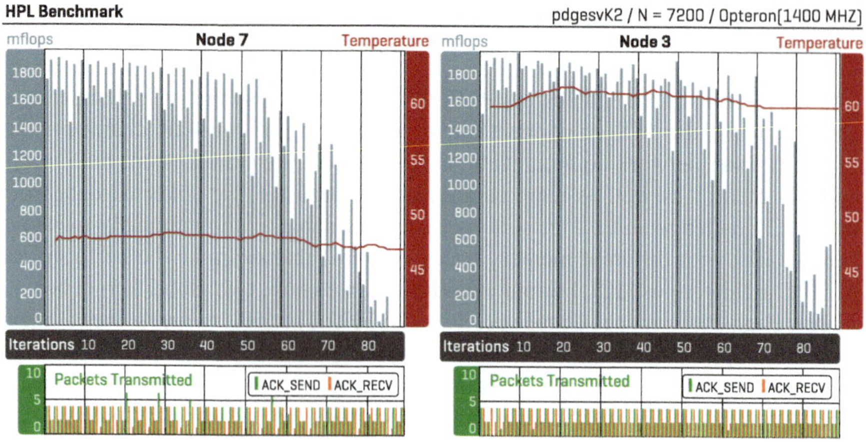

Fig. 11.2: FLOPS, temperature and communication monitoring using the CPU, ACPI and MX component of PAPI-C for an HPL benchmark running on an AMD Opteron cluster

11.2.7 The LM-SENSORS Component

The LM-SENSORS component enables the PAPI-C library to access all computer health monitoring sensors as exposed by the `lm_sensors` [13] library. The user is able to closely monitor the system's hardware health as an attempt to get more performance out of environmental conditions of the hardware. What features are available and what exactly can be monitored depends on the hardware setup.

We monitored three fan speeds as well as the CPU temperatures on a quad-core Intel Nehalem (core i7) machine using the LM-SENSORS component of PAPI-C. Multiple iterations of numeric operations are performed to heat up the compute cores. In total, 128 threads have been created and distributed over 8 compute cores and each of them executes the numeric code. The fan speeds as well as CPU temperatures are monitored every 10 seconds. Figure 11.3(a) shows the collected speed data of three fans while Fig. 11.3(b) depicts the temperature of the two quad-core CPUs. From those graphs, it is evident that the rotational speed of the fans responds to changes on the CPU temperature sensors. Note once more the difference in temperature between the two CPUs. We have seen similar correlation between temperature and workload before in section 11.2.6 on an Opteron architecture.

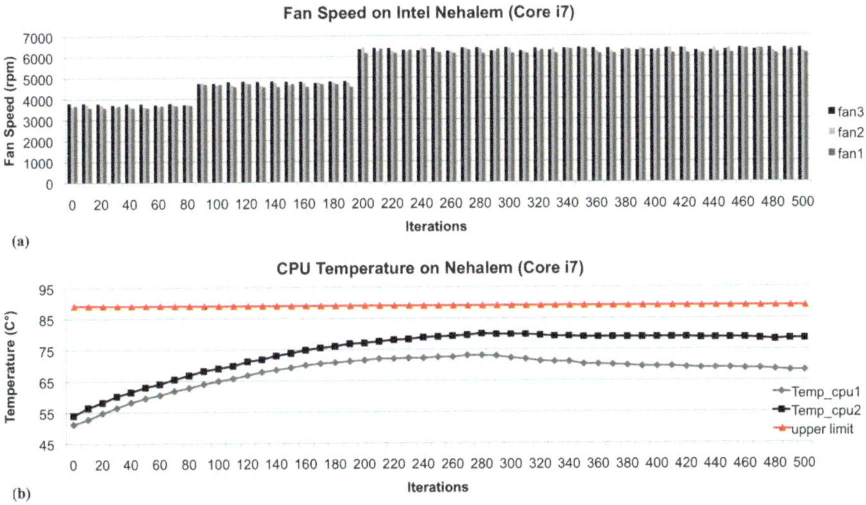

Fig. 11.3: (a) Fan speed monitoring; (b) CPU temperature monitoring - both metrics have been investigated on an Intel Nehalem (core i7) machine using the LM-SENSORS component of PAPI-C

11.3 Multi-core Performance Measurement

With the arrival of the multi-core era for modern Petascale computing, more discussions are turning to the future implications of multi-core processors. The main focus in this section is the impact of shared resources of multi-core processors on the CPU component of PAPI-C which is described in 11.2.4. With the help of an application test case, we will discuss the difference between hardware performance data collection for on-core versus off-core resources. The current approach of collecting hardware performance counters shows serious limitations for off-core resources. However, measurement of performance counter data from shared resources is crucial in the analysis of scientific applications on multi-core processors due to the fact that this is where resource contention occurs. The key is to minimize the contention of shared resources such as caches, memory bandwidth, bus and other resources.

The multi-core transition in hardware design also reflects an impact on software development which remains a big challenge. To illustrate issues associated with the measurement of performance events for shared resources, we quantitatively evaluate the performance of the memory sub-system on Jaguar, the fastest computer on the November 2009 Top500 list [14]. The Jaguar system at Oak Ridge National Laboratory (ORNL) has evolved rapidly over the last several years. When the work reported here was done, Jaguar was based on Cray XT4 hardware and utilized 7,832 quad-core AMD Opteron processors with a clock frequency of 2.1 GHz and 8 GBytes of memory (maintaining the per core memory at 2 GBytes). For more information on the Jaguar system and the quad-core AMD Opteron processor, the reader is referred to [15, 16].

The application test case is drawn from workload configurations that are expected to scale to large number of cores and that are representative of Petascale problem configurations. The massively parallel direct numerical simulation (DNS) solver (S3D) - developed at Sandia National Laboratories - solves the full compressible Navier-Stokes, total energy, species, and mass continuity equations coupled with detailed chemistry [17, 18, 19]. The application was run in SMP (one core per node) as well as VN mode (four cores per node) on Jaguar. Both test cases apply the same core count. The total execution time for runs using the two different modes shows a significant slowdown of 25% in VN mode (813 seconds) when compared to single-core mode (613.4 seconds). The unified L3 cache is shared between all four cores. We collected hardware performance events using the PAPI library that confirms our findings. L3 cache requests are measured and computed using the following PAPI native events:

```
L3 REQUESTS = READ REQUESTS TO L3 + L3 FILLS CAUSED BY L2 EVICTION
```

Note: In VNM all L3 cache measurements have been divided by 4 (4 cores per node)

Figure 11.4 (a) depicts the number of L3 cache misses and requests when using four cores versus one core per node for the 13 most expensive functions of the S3D application. It appears that the performance degradation in VN mode is due to the L3 cache behavior. In VN mode we see roughly twice as many L3 cache requests and misses compared to SMP mode. It is not surprising that L3 cache misses increase with VN mode since if every thread is operating on different data, then one thread could easily evict the data for another thread if the sum of the four working threads is greater than the size of the L3 cache. However, the increase in L3 requests is rather questionable. The L3 cache serves as a victim cache for L2. In other words, a datum evicted from L2 (the victim) is deposited in L3. If requested data is not in L2 cache then the L3 cache is checked which results in an L3 request. While the L3 cache is shared between all four cores, the L2 cache remains private. Based on this workflow, it is not clear why the number of L3 requests increases so dramatically when using all four cores per node. As verification we measure the L2 cache misses in SMP and VN mode and Fig. 11.4 (b) presents the comparison. It clearly shows that the number of L2 cache misses does not increase when all four cores are used compared to SMP mode. All the more, the question persists as to where the double L3 cache requests come from when VN mode is used. It is important to note, the policy on the Jaguar system defines that by default a task - independent of process or thread - is not allowed to migrate to a CPU core within a socket or to any CPU core on either socket [20]. For the S3D test case, we applied this default configuration which pins a task to a specific CPU core.

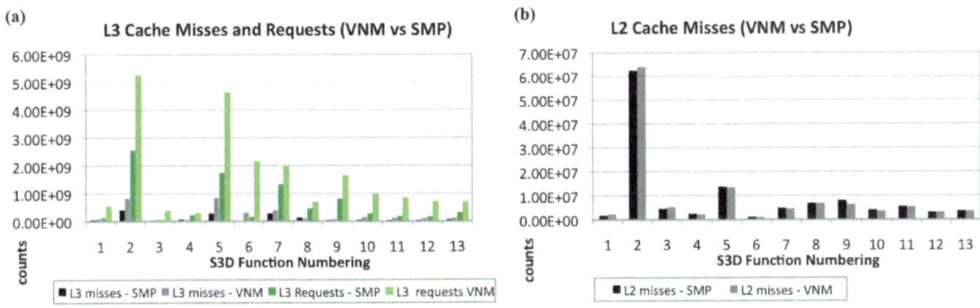

Fig. 11.4: (a) L3 cache misses and requests (mean); (b) L2 cache misses (mean)

11.3.1 Various Multi-core Designs

Recent investigations and discussions have suggested that the high L3 cache request rate in S3D may be an artifact of the measurement process. Current Opteron hardware is not designed for first-person counting of events involving shared resources [21]. The L3 events in AMD Opteron quad-core processors are not monitored in four independent sets of hardware performance registers but in a single set of registers not associated with a specific core (often referred to as "shadow" registers). Each core has four independent counter registers which are used for most performance events. When an L3 event is programmed into one of these counters on one of these cores, it gets copied by hardware to the shadow register. Thus, only the last event to be programmed into any core is the one actually measured by all cores. When several cores try to share a shadow register, the results are not clearly defined. Performance counter measurement at the process or thread level relies on the assumption that counter resources can be isolated to a single thread of execution. That assumption is generally no longer true for resources shared between cores - like the L3 cache in AMD quad-core processors.

This problem is not isolated just to AMD Opteron processors. Early Intel dual-core processors addressed the issue of measuring shared resources by providing SELF and BOTH modifiers on events involving shared caches or other resources. This allowed each core to independently monitor the event stream for a shared resource and to either collect only counts for its activity or for all activities involving that resource. However, with the introduction of the Nehalem (core i7) architecture, Intel, too, moved measurement of chip level shared resources off the cores and onto the chips. The Nehalem architecture includes eight "Uncore" counters [22] that are shared among all the cores of the chip. There is presently no mechanism for a given core to reserve counter resources from the Uncore. These events can be monitored by the `perfmon2` [5] patch, but only in system-wide counting mode. Thus these counter measurements cannot be performed with a first-person measurement paradigm such as PAPI's, and cannot be intermixed with per process measurements of other events. The built-in `perf_events` [6] module in the Linux kernel has no support for Uncore counters as of the 2.6.32 kernel release.

A final example of the multi-core problem of measuring activities on shared resource is IBM's Blue Gene series. Blue Gene/L is a dual-core processor and Blue Gene/P is a quad-core processor. In both cases hardware counters are implemented in the UPC, a Universal Performance Counter module that is completely external to any core. In Blue Gene/P for example, the UPC contains 256 independent hardware counters [23]. Events on each core can be measured independently, but the core must be specified in the event identifier. This can create great difficulty for code that in general does not know or care on which core it is running. Further, these counters can only be programmed to measure events on either core 0 and 1, or core 2 and 3, but not on a mixture of all four cores at once.

As the above examples illustrate, hardware vendors are searching for ways to provide access to performance events on shared resources. There is presently no standard mechanism that provides performance information in a way that is useful for software tuning. New methods need to be developed to appropriately collect and interpret hardware performance counter information collected from such multi-core systems with interesting shared resources. PAPI research is underway to explore these issues.

11.4 Future Directions

With the release of PAPI-C, the stage is set for a wide range of development directions. Our major goals with the first release were stability and compatibility. As with any research and development effort there are always open issues to be explored. Here are some of the issue under investigation with Component PAPI:

- **Event Naming:** PAPI presently expresses all events as 32-bit event codes. With the richness of current events and attributes and modifiers, we find this too restrictive, and will be migrating to a model in which all events are referenced by name.
- **Data Types:** PAPI supports returned data values expressed as unsigned 64-bit integers. This is appropriate for counting events, but may not be as appropriate for expressing other values. We are exploring ways to encode and specify other 64-bit data formats including: signed integer, IEEE double precision, fixed point, and integer ratios.
- **Dynamic Configurability:** The current mechanism for adding new components is workable, but not well suited to introducing new components between releases of the PAPI Framework. Methods are needed for an automated discovery process for components, both at build time and at execution time.
- **Synchronization:** Components can report values with widely different time scales and remote measurements may exhibit significant skew and drift in time from local measurements. Mechanisms need to be developed to accomodate these artifacts.

- **Component Management:** To encourage users and third parties to become component contributors, efforts will be invested in documenting the component development process and in managing 3rd party components.

At a recent brainstorming session by the PAPI developers, a number of future directions for the PAPI project were identified. In a somewhat whimsical fashion, and building on the idea of the PAPI-C name, several new letters for the PAPI "alphabet soup" were put forth:

- **PAPI-M: Multi-core.** The issue of how to measure shared resource performance on a variety of multi-core architectures remains unresolved. This may require more kernel development than PAPI development, but is an important issue that should be addressed.
- **PAPI-G: GPUs.** GPGPUs and other heterogenous compute elements will be an increasingly important part of our computing eco-system as we move from Petascale to Exascale. They present radically different sorts of performance information to the user and provide a challenging opportunity for performance presentation.
- **PAPI-V: Virtual.** With access to performance hardware now part of the Linux kernel, it becomes possible to introduce this information into the hypervisors that comprise virtual, or cloud, computing space. With support in the hypervisors, it becomes possible to consider what it means to measure hardware performance in the cloud.
- **PAPI-N: Networks.** As core counts rise exponentially on the march to Exascale, communication becomes even more dominant over computation as a determinant of execution time. PAPI-C components can be developed either in the open source community or by vendors to monitor hardware characteristics of either open network standards such as Infiniband or proprietary hardware such as Cray's SeaStar or Gemini network chips.
- **PAPI-D: Disks.** Several users of PAPI have suggested and begun work on the development of PAPI Components to measure remote disk storage activities for file systems like Lustre. Such information could prove useful in managing and measuring the impact of storage operations on execution performance.
- **PAPI-H: Health.** System health measurements are often done out-of-band from compute activities. PAPI-C components may be developed to run on system nodes in parallel with jobs on compute nodes to assess the impact of application activities on temperature or power consumption, or to warn of impending resource failure and the need for remedial action.

11.5 Conclusion

For most of the past decade, PAPI has been the de-facto choice to provide the tool designer and application engineer with a consistent interface for accessing hardware performance counters on a wide range of computer architectures. PAPI has ridden the evolutionary wave of processor development as clock rates, pipeline depth and instruction level parallelism increased through the decade. That smooth evolution has recently ended with the flattening of clock rates, the introduction of multi-core architectures, the adoption of heterogeneous computing approaches and the need for more careful monitoring of system health required for fault tolerance and resiliency in the Petascale domain of hundreds of thousands of processors. We are now in a period of punctuated equilibrium where the paradigms of the recent past are being swept away by a tidal wave of changes at a number of levels.

The development of Component PAPI for the simultaneous monitoring of multiple measurement domains positions this library to remain as a central tool in the acquisition of performance data across a spectrum of architectures and activities. This extension has been done in such a way as to cause minimal disruption to the current user base while providing flexible opportunities to gain new insights into application and system performance.

Acknowledgements This research was sponsored in part by the Office of Mathematical, Information, and Computational Sciences of the Office of Science (OoS), U.S. Department of Energy (DoE), under Contract No. DE-AC05-00OR22725 with UT-Battelle, LLC. This work used resources of the National Center for Computational Sciences at Oak Ridge National Laboratory, which is supported by the Office of Science of the Department of Energy under Contract DE-AC05-00OR22725. These resources were made available via the Performance Evaluation and Analysis Consortium End Station, a Department of Energy INCITE project.

This work was also supported in part by the U.S. Department of Energy Office of Science under contract DE-FC02-06ER25761, by the National Science Foundation, Software Development for Cyberinfrastructure (SDCI) Grant No. NSF OCI-0722072 Subcontract No. 207401, and by the Department of Defense, using resources at the Extreme Scale Systems Center.

References

1. Browne, S., Dongarra, J., Garner, N., Ho, G., Mucci, P.: A portable programming interface for performance evaluation on modern processors. International Journal of High-Performance Computing Applications, Vol. 14, No. 3, pp. 189-204 (2000)
2. Cameron, K.W., Ge, R., and Feng, X.: High-performance, power-aware distributed computing for scientific applications. Computer, 38(11):40–47 (2005)
3. Feng, W.C.: The importance of being low power in high performance computing. CTWatch Quarterly, 1(3), August (2005)
4. Freeh, V.W., Lowenthal, D.K., Pan, F., Kappiah, N.: Using multiple energy gears in MPI programs on a power-scalable cluster. In Principles and Practices of Parallel Programming (PPOPP), June (2005)
5. `Perfmon2` Sourceforge Project Page: `http://perfmon2.sourceforge.net`
6. Molnar, I.: Performance Counters for Linux, v8. `http://lwn.net/Articles/336542`
7. Moore, S.: A Comparison of Counting and Sampling Modes of Using Performance Monitoring Hardware. ICCS 2002, Amsterdam, April (2002)
8. Operating System share, November 1999: `http://www.top500.org/charts/list/14/os`
9. Operating System share, November 2009: `http://www.top500.org/charts/list/34/os`

10. Pettersson, M.: Linux x86 Performance-Monitoring Counters Driver. http://www.csd.uu.se/~mikpe/linux/perfctr
11. Jarp, S., Jurga, R., Nowak, A.: Perfmon2: A leap forward in Performance Monitoring. Journal of Physics: Conference Series 119, 042017 (2008)
12. Luszczek, P., Dongarra, J., Koester, D., Rabenseifner, R., Lucas, B., Kepner, J., McCalpin, J., Bailey, D., Takahashi, D.: Introduction to the hpc challenge benchmark suite. Technical report, March (2005)
13. Hardware Monitoring by lm_sensors: http://www.lm-sensors.org/
14. Top500 list: http://www.top500.org
15. NCCS.gov computing resources documentation: http://www.nccs.gov/computing-resources/jaguar
16. Software Optimization Guide for AMD Family 10h Processors, Pub. no. 40546 (2008)
17. Chen, J. H., Hawkes, E. R., et al.: Direct numerical simulation of ignition front propagation in a constant volume with temperature inhomogeneities I. fundamental analysis and diagnostics. Combustion and flame, 145, pp. 128-144 (2006)
18. Sankaran, R., Hawkes, E. R., et al.: Structure of a spatially developing turbulent lean methane-air Bunsen flame. Proceedings of the combustion institute 31, pp. 1291-1298 (2007)
19. Hawkes, E. R., Sankaran, R., et al.: Scalar mixing in direct numerical simulations of temporally evolving nonpremixed plane jet flames with skeletal CO-H2 kinetics. Proceedings of the combustion institute 31, pp. 1633-1640 (2007)
20. Cray XT Programming Environment User's Guide (Version 2.2). S-2396-22, July (2009)
21. BIOS and Kernel Developer's Guide (BKDG) for AMD Family 10h Processors (particularly Section 3.12.). Vol. 31116 Rev 3.34, September (2009)
22. Intel 64 and IA-32 Architectures Software Developer's Manual Volume 3B: System Programming Guide (Particularly Chapter 19.17.2 Performance Monitoring Facility in the Uncore). Part 2 Order Number: 253669-031US, June (2009)
23. Walkup, B.: Blue Gene/P Universal Performance Counters. http://www.nccs.gov/wp-content/training/2008_bluegene/BobWalkup_BGP_UPC.pdf

Chapter 12
ISP Tool Update: Scalable MPI Verification

Anh Vo, Sarvani Vakkalanka, and Ganesh Gopalakrishnan

Abstract We provide a status update of ISP, our dynamic formal verifier for MPI programs. ISP determines and explores all relevant schedules of an MPI program. The highlights of this paper are (i) a recap of ISP's features, and (ii) an overview of work in progress, including a graphical explorer for message passing (GEM) and a distributed MPI analyzer (DMA), an adaptation of ISP for the distributed setting.

12.1 Introduction

The Message Passing Interface (MPI) [6] library remains one of the most widely used APIs for implementing distributed message passing programs. Its projected usage in critical, future applications such as Petascale computing [5] makes it imperative that MPI programs be free of programming logic bugs. This is a very challenging task considering the size and complexity of optimized MPI programs. In particular, performance optimizations often introduce many types of non-determinism in the code. For example, the `MPI_Recv(MPI_ANY_SOURCE, MPI_ANY_TAG)` call that can potentially match a message from any sender in the same communication group (we will later refer to this as a *wildcard receive*) is often used for re-initiating more work on the first sender that finishes the previous item of work. A more general version of this call is the `MPI_Waitsome` call that waits for a subset of the previously issued communication requests to finish. These non-deterministic constructs result in MPI program bugs that manifest intermittently – the bane of debugging.

Traditional MPI debugging tools such as Marmot [8] insert delays during repeated testing under the same input to pertube the MPI runtime scheduling. Experience indicates that this technique is often unreliable [1]. In order to detect all scheduling-related bugs, the framework under which MPI programs are debugged

Anh Vo, Sarvani Vakkalanka, Ganesh Gopalakrishnan
School of Computing, University of Utah, Salt Lake City, UT 84112, USA

M.S. Müller et al. (eds.), *Tools for High Performance Computing 2009*,
DOI 10.1007/978-3-642-11261-4_12, © Springer-Verlag Berlin Heidelberg 2010

needs to have the ability to *determine* and *enforce* all *relevant* schedules. These three concepts are explained in § 12.1.1, including why only the relevant schedules must be enforced. These concepts underlie ISP [11, 12, 13, 14], a unique dynamic verifier for MPI programs. In the rest of this paper, we provide an updated status of ISP matching the tool presentation in the workshop.

12.1.1 Determining and Enforcing Relevant Schedules

Determining Schedules: In an MPI program, the MPI calls are always issued in program order, and the computational (C or Fortran) code is also executed in program order. However, MPI calls may match out of program order [12]. Furthermore, there are non-deterministic send/wild-card receive matches that cannot be discovered unless these out of order matches are considered.

```
P0:Isend(to P1, &h0) ; Barrier;                          Wait(h0);
P1:Irecv(*, &h1)     ; Barrier;                          Wait(h1);
P2:                   Barrier; Isend(to P1, &h2);        Wait(h2);
```

Fig. 12.1: Illustration of Barrier Semantics and the POE Algorithm

Consider the example in Figure 12.1 where an `MPI_Isend` is issued by `P0` and another `MPI_Isend` is issued by `P2`, both directed at `P1` which issues a wildcard `Irecv`. There are `Barrier` calls intersperced in this code.

According to the MPI library semantics, no MPI process can *issue* an instruction past its barrier unless all other processes have *issued* their barrier calls; however these calls need not be matched when a Barrier is crossed. The following execution is possible: (i) `Isend(to P1, h0)` is issued but not matched yet, (ii) `Irecv(*, h1)` is issued but not matched, (iii) the processes issue *and match* their barriers, (iv) `Isend(to P1, h2)` is issued but not matched, and (v) now both sends and the receive are alive, and non-determinism arises.

However, notice that in any schedule where P0's `Isend` and P1's `Irecv` are forced to match (*i.e.*, issue order matchings), there is no non-determinism in this program! ISP is the only dynamic formal verification tool into which the out of order matching semantics of MPI is built in.

Enforcing Schedules: Once we determine that non-determinism is possible in Figure 12.1, we need to have a scheduling strategy to discover this non-determinism and then to enforce it. ISP discovers non-determinism through a single idea (basically): *delay any non-blocking non-deterministic receive if allowed by the MPI semantics* [12]!

In this example, the ISP scheduler collects the `Irecv` command but does not issue it into the MPI runtime (effectively delaying it). This forces the execution to proceed in such a manner that the non-deterministic matches are revealed.

Once revealed, ISP invokes its program replay mechanism. That is, ISP is designed to remember the non-deterministic choices discovered and re-execute an MPI program to cover this space of non-deterministic matches. During each *replay*, it *dynamically rewrites* the wildcard `Irecv(*, &h1)`: first, it is turned into a `Irecv(from P0, &h1)` and during the next replay, it is turned into `Irecv(from P2, &h1)`. This way, the ISP scheduler can "fire and forget" the `Irecv(from P0, &h1)` and `Isend(to P1, &h0)`, knowing that within the MPI runtime, these instructions have no choice, but to match! Had the ISP scheduler simply issued `Irecv(*, &h1)` (without determinizing it first), this receive could have matched *a send from yet another process, say P3* – not the intended one.

Why only the *Relevant* Schedules? If *all possible schedules* of an MPI program are considered, a considerable degree of wasted testing happens. For example, nothing is gained by permuting the order in which (i) the send/receive MPI calls are *issued*, (ii) the order in which the `Barrier's` are *issued*. For instance, a delay introducing tool may simply explore all $N!$ ways of issuing a `Barrier` call whereas ISP simply issues the barriers in one canonical way (it issues them together).

Thus the key difference between an *ad hoc* testing tool and a formal dynamic verifier such as ISP is that whereas the former tools get caught in an exponentially growing amount of irrelevant schedules, ISP avoids *many* of these exponentials. In practice, this makes all the difference between complete verification and ad hoc testing with omissions. For example, in a recent series of dynamic verification runs using ISP, ParMETIS [7] (a widely used hypergraph partitioner of over 14,000 lines of MPI C code) was verified [2] for 32 processes on an ordinary laptop computer in under 40 minutes, generating a message log file of 512 MB! This verification could simply not be imagined if we permuted all 32-way barriers within this code, posted all determinstic sends and receives in every order, etc.

12.2 ISP Overview

At a high level, ISP works by intercepting the MPI calls made by the target program and making decisions on when to send these MPI calls to the MPI library. This is accomplished by the two main components of ISP: the Interposition Layer and the Scheduler. Figure 12.2 describes the general framework of ISP, showing the interaction between the ISP library (linked with the executable) and the ISP scheduler.

The Interposition Layer: The interception of MPI calls is accomplished by compiling the ISP interposition layer together with the target program source code. The interposition layer makes use of the profiling mechanism PMPI. It provides its own version of `MPI_`f for each corresponding MPI function f. Within each of these

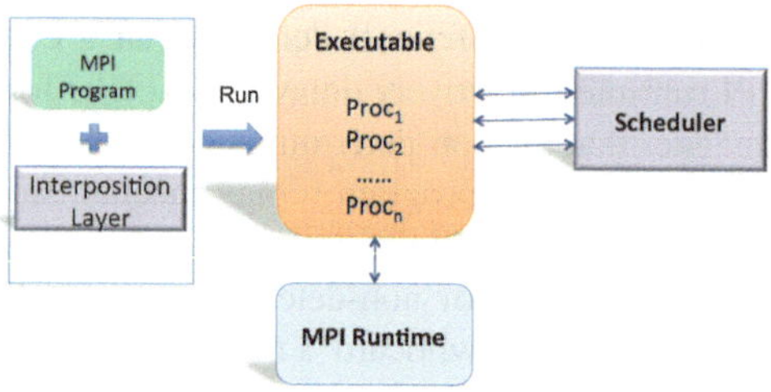

Fig. 12.2: Overview of ISP

`MPI_`*f*, the profiler communicates with the scheduler using TCP sockets[1] to send information about the MPI call that the process wishes to make. It will then wait for the scheduler to make a decision on whether to send the MPI call to the MPI library or to postpone it until later. When permission to fire *f* is granted from the scheduler, the corresponding `MPI_`*f* will be issued to the MPI run-time. Since all MPI libraries come with functions such as *PMPI_f* for every MPI function *f*, this approach provides a portable and light-weight instrumentation mechanism for MPI programs being verified.

The ISP Scheduler: The scheduler is where our main scheduling algorithm, namely POE (Partial Order avoiding Elusive interleavings) is carried out. The scheduler meets the following objectives: G1: discovers the maximal set of sends that can match a wildcard receive (viewed across all MPI-standard compliant MPI libraries); G2: accurately models the semantics of the global operations (such as barriers) of MPI. In MPI, not all MPI operations issued by a process complete in that order, and a proper modeling of this out-of-order completion semantics is essential in order to meet goals G1 and G2. For example, two `MPI_Isend` commands issued in succession by an MPI process P1 to the same target process (say P2) are forced to match in order, whereas if these `MPI_Isends` are targeted to two *different* MPI processes, then they *may match contrary to the issue order*. As another example, any operation following an `MPI_Barrier` must complete only after the barrier has completed, while an operation issued *before* the barrier may linger across the barrier, as already illustrated using Figure 12.1.

Main Steps of the POE Algorithm: The POE algorithm works as follows. There are two classes of statements to be executed: (i) those statements of the embedding programming language (C, C++, and Fortran) that do not invoke MPI commands and (ii) the MPI function calls. The embedding statements in an MPI process are local in the sense that they have no interactions with those of another process. Hence, under POE, they are executed in program order. When an MPI call *f* is encountered, the scheduler records it in its state; however, it does not (necessarily) issue this call into the MPI run-time. (Note: When we say that the scheduler issues/executes MPI call *f*, we mean that the scheduler grants permission to the process to issue the corresponding PMPI_*f* call to the MPI run-time). This process continues until the

[1] When running within a local machine, ISP uses unix sockets to reduce communication overhead.

scheduler arrives at a *fence*, where a fence is defined as an MPI operation that cannot complete after any other MPI operation following it. The list of such fences include `MPI_Wait`, `MPI_Barrier`, *etc.*, and are formally defined in [11]. When all MPI processes are at their individual fences, the full extent of all senders that can match a wildcard receive becomes known, and dynamic rewriting can be performed with respect to these senders. The collection of sends and matching receives can then be issued. In the case where a receive does not have a matching send (or vice versa) and there is no other MPI calls that can proceed, a deadlock is declared by the scheduler.

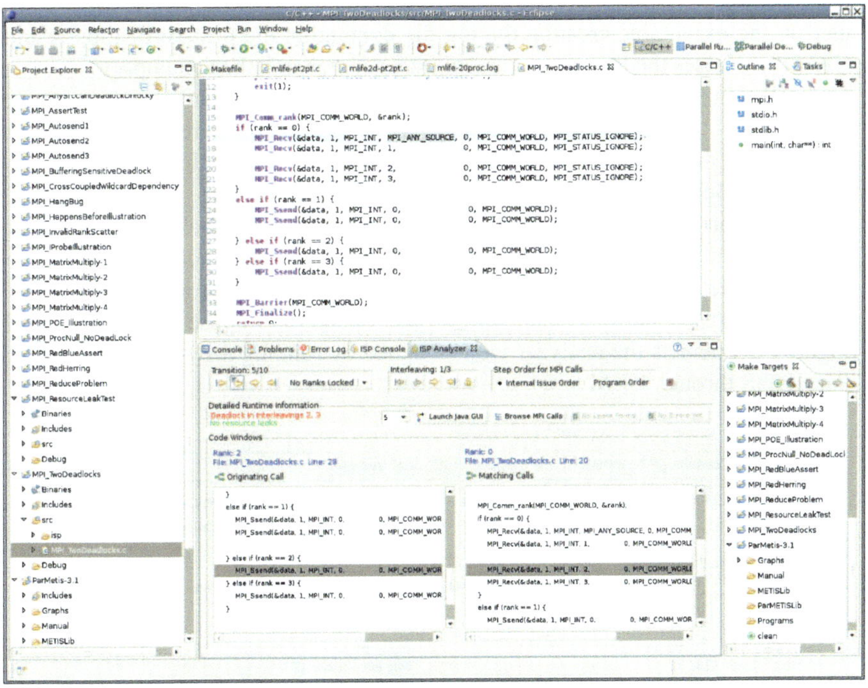

Fig. 12.3: ISP plugin in Eclipse

Completes-Before Ordering: The Completes-Before (CB) ordering accurately captures when two MPI operations x and y issued from the same process in program order are guaranteed to complete in that order. For example, if an MPI process `P1` issues an `MPI_Isend` that ships a large message to `P2` and then issues `MPI_Isend` that ships a small message to `P3`, it is possible for the second `MPI_Isend` to complete first. A summary of the completes-before order of MPI is as follows: (i) **Send Order**: Two `Isends` sending data to the same destination complete in issue order. (ii) **Receive Order**: Two `Irecvs` receiving data from the same source complete in issue order. (iii) **Wildcard Receive Order**: If a wildcard `Irecv` is followed by another `Irecv` (wildcard or not), the issue order is respected by the completion order.

(iv) **Wait Order**: A `Wait` and another MPI operation following it complete in issue order. For a formal description of the CB relation, please see [12].

12.3 GEM

GEM (Graphical Explorer for Message passing) is an Eclipse plugin which serves as a graphical front-end for ISP. Given a collection of files to analyze using ISP, GEM helps compile and links the files using the ISP library, and then invoke ISP's scheduler on the executable creating the log file containing the post-verifcation results. GEM then parses the log file and organizes its contents. It then attempts to associate MPI calls with one another (e.g., sends need to be associated with their corresponding receives). GEM also allows users to view the execution results according to the program order or according to ISP's internal execution order (modeled by the Completes-Before-Ordering).

In addition to the usual Eclipse textual console view, GEM also provides an analyzer view that serves three functions: (i) summarize verification results, (ii) link to the completes-before viewer, (iii) allow the users to step through matching MPI calls. Figure 12.3 shows a small MPI project opened in Eclipse under GEM while Figure 12.4 depicts the analyzer view obtained by running a 10-process version of ParMETIS through GEM.

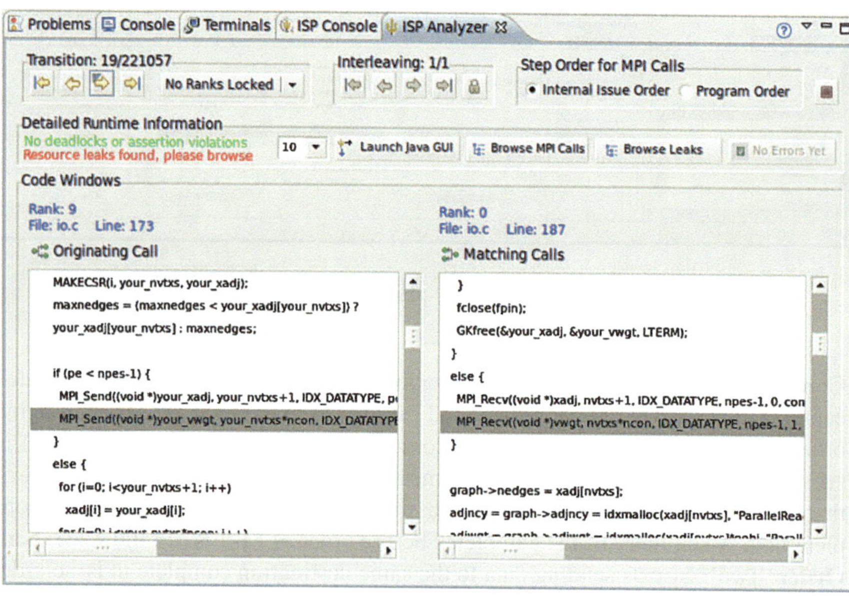

Fig. 12.4: ISP plugin in Eclipse

For a more detail description of GEM, please see [2].

12.4 DMA - A Distributed MPI Analyzer

Up until now, ISP has been used to verify several large programs such as ParMETIS, MADRE [10], MPI-Blast [3], and some of the SPECMPI2007 [4] benchmarks. However, most of the experiments have been confined a small number of processes. Fortunately, most MPI bugs will manifest themselves within these scale-down experiments. Unfortunately, *some* MPI bugs can only be reproduced when the program is run with a large number of processes. With Petascale computing coming to reality and Exascale computing looming in the horizon, the need for a scalable tool that can formally verify very large MPI programs is clear. § 12.4.1 will discuss several technical limitations of ISP and we will propose our design of the new framework in § 12.4.2.

12.4.1 Limitations of ISP

ISP was designed as a debugger that works best in a single machine setting. The assumption is that most MPI bugs still manifest themselves even with a small number of processes. However, as mentioned before, there are bugs that will manifest only at the extreme end of computing and it is unrealistic to expect the developers to roll back to a desktop version of the program to debug. Since [13] was published, many improvements were made to ISP to enhance the performance of the tool, both in terms of scalibity and speed. However, there are key technical limitations of ISP which prevent further significant scaling. We list the two most important issues here: **Centralized Scheduler:** ISP employs a centralized scheduler which communicates with the MPI processes through TCP sockets. Later versions of ISP can take advantage of Unix sockets when operating within a machine to eliminate the socket communication overhead. For small to medium MPI programs, the verification can complete within a reasonable amount of time in a desktop/workstation. However, for large programs, the MPI processes will often need to be executed in a cluster environment (launching a couple hundred threads on a local machine can slow down the verification tremendously, even on today's powerful machines). While ISP can operate in distributed mode, which means that the processes compiled with the interposition layer can be run distributedly in a cluster, the scheduler remains centralized and becomes the bottleneck of the whole system. This is mainly due to amount of sequential processing that has to be done by the scheduler, and also due to the memory limitation of a single machine. Despite being a stateless verifier (meaning that it does not store all visited states), ISP still needs to store the information along the current execution trace so that it can remember correctly which interleavings to

explore. When the number of MPI processes and MPI calls become large, even this space grows inordinately.
Synchronous Communication: The ISP scheduler and the MPI processes operates in a lock-step fashion, which means the MPI processes communicate in a handshake fashion with the scheduler about each MPI call (except those that have purely local semantics such as type creation, communicator grouping, etc.). Since ISP has to hold back communication to build the completes-before edges properly, this handshake mechanism is necessary. However, when the number of processes is large enough, the scheduler has to cycle through each process to do handshaking, which means several MPI processes will unnecessarily block to wait for the response from the ISP scheduler.

12.4.2 DMA

In order to address the limitations of ISP, we have designed a new dynamic verification framework called DMA. Since the implementation is still in its early phase, this paper will only provide a brief overview of the framework.

12.4.2.1 PnMPI

DMA is designed to operate as a P^NMPI [9] module. P^NMPI extends the PMPI profiling interface to support multiple PMPI-based tools (ISP is an example of a PMPI-based tool). The advantages of using P^NMPI are as follows:

- P^NMPI allows the implementation of the DMA tool to be split up into multiple layers, with each layer addressing orthogonal issues (interleaving generation layer, resource leak tracking layer, deadlock detection layer, etc.).
- P^NMPI also eliminates the need to recompile the MPI target code every time changes are made to DMA
- Each layer or all of DMA layers can be turned off through a simple configuration file, which enables the developers quickly to choose which aspect of the program he wants to debug (or not debug at all)
- P^NMPI has very low overhead. For more details on P^NMPI overhead, please see [9]

12.4.2.2 Going Distributed

With the lessons learned from developing ISP, DMA is designed as a distributed tool. In DMA, each process will maintain its own trace and figure out the *potential* matching that can occur for each nondeterministic MPI event that happens within itself. To accomplish this, each process needs to construct its *view of the world* through *piggybacking* (i.e., sending extra information in each MPI message), which itself is implemented as a P^NMPI module. Intuitively, the view helps the process reason about the completes-before relationship that was described earlier. In the case of ISP, the scheduler maintains the trace record of all the processes and thus can construct the global view by itself. DMA maintains the trace record in each process and eliminates the need for centralized processing.

After the program finishes, the information written out by each process can then be processed offline by a *Scheduler Generator* which then generates the necessary interleavings upon which the MPI processes will follow upon restarting. During all these interleavings exploration, other DMA layers can also check the processes for resource leaks, deadlocks, and local assertion violations.

12.4.2.3 Practical Considerations

DMA currently has two different implementations which serve different set of programs. The first implementation ignores the potential dependency between concurrent non-deterministic events, in which certain choices made by one event can affect the number of choices for the other event. The second implementation takes into account all the possible dependencies between concurrent events. Obviously, the second implementation has a higher overhead than the first one. However, based on our experiments with ISP, many programs do not exhibit dependency between concurrent nondeterministic events.

In our experience, many large MPI programs tend to follow several programming patterns, such as master-slave, work-stealing, nearest-neighbor, etc. Several of these patterns employ heavy usage of wildcard receives, which results in an exponentially large number of relevant interleavings to be explored. Yet, only a handful of these interleavings represent meaningful code paths that have to be verified. The rest of them can be declared safe once a representative of those interleavings is checked to be safe. The analysis for this is complex and is part of future work for DMA

12.5 Conclusions

In this paper, we provided a status update of our dynamic verifier ISP matching the tool presentation made at the 3rd Parallel Tools Workshop. We described the graphical integration (fully explained in [2]) and a distributed version of ISP under design.

Our future plans include a true *in situ* verification of large MPI programs made possible by the distributed dynamic verification algorithm of the DMA tool. Our plans also include an integration of the graphical capabilities of the tool presented at [2] with DMA.

Acknowledgements The authors have a long list of Gauss group members as well as researchers of the Lawrence Livermore National Laboratories (LLNL) as well as IBM to thank: of the Gauss group members, we like to especially point out the work of Alan Humphrey and Chris Derrick of Utah on the GEM tool, in collaboration with Beth Tibbitts and Greg Watson (IBM). We also thank Bronis de Supinski, Martin Schulz, and Greg Bronevetsky of LLNL for joint work on DMA.

References

1. `http://www.cs.utah.edu/formal_verification/ISP_Tests/`.
2. `http://www.cs.utah.edu/formal_verification/ISP-Eclipse/`.
3. `http://www.mpiblast.org/`.
4. `http://www.spec.org/mpi`.
5. A. Geist. Sustained Petascale: The next MPI challenge. Invited Talk at EuroPVM/MPI 2007.
6. W. Gropp, E. Lusk, N. Doss, and A. Skjellum. A high-performance, portable implementation of the MPI message passing interface standard. *Parallel Computing*, 22(6):789–828, Sept. 1996.
7. G. Karypis. METIS and ParMETIS. `http://glaros.dtc.umn.edu/gkhome/views/metis`.
8. B. Krammer, K. Bidmon, M. S. Müller, and M. M. Resch. Marmot: An MPI analysis and checking tool. In *Parallel Computing 2003*, Sept. 2003.
9. M. Schulz and B. R. de Supinski. P^NMPI tools: A whole lot greater than the sum of their parts. In *SC*, page 30, 2007.
10. S. F. Siegel and A. R. Siegel. MADRE: The Memory-Aware Data Redistribution Engine. EuroPVM/MPI 2008
11. S. Vakkalanka, G. Gopalakrishnan, and R. M. Kirby. Dynamic verification of MPI programs with reductions in presence of split operations and relaxed orderings, *CAV* 2008.
12. S. Vakkalanka, A. Vo, G. Gopalakrishnan, and R. M. Kirby. Reduced execution semantics of MPI: From theory to practice. In *FM*, pages 724–740, 2009.
13. A. Vo, S. Vakkalanka, M. DeLisi, G. Gopalakrishnan, R. M. Kirby, and R. Thakur. Formal verification of practical MPI programs. In *PPoPP*, pages 261–269, 2009.
14. A. Vo, S. S. Vakkalanka, J. Williams, G. Gopalakrishnan, R. M. Kirby, and R. Thakur. Sound and efficient dynamic verification of MPI programs with probe non-determinism. In *EuroPVM/MPI*, pages 271–281, 2009.

GPSR Compliance

The European Union's (EU) General Product Safety Regulation (GPSR) is a set of rules that requires consumer products to be safe and our obligations to ensure this.

If you have any concerns about our products, you can contact us on ProductSafety@springernature.com

In case Publisher is established outside the EU, the EU authorized representative is:

Springer Nature Customer Service Center GmbH
Europaplatz 3
69115 Heidelberg, Germany

Batch number: 10585138

Printed by Printforce, the Netherlands